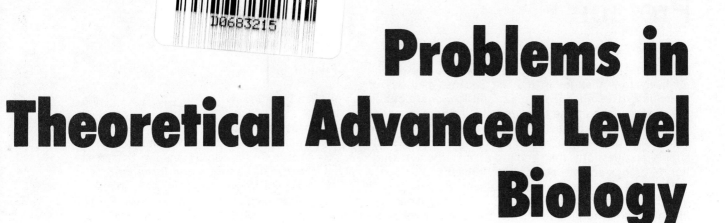

Problems in Theoretical Advanced Level Biology

P.W. Freeland

B.Sc., M. Phil., Dip. Ed., F.I. Biol.

Head of Science, Worth School, Sussex

HODDER AND STOUGHTON

LONDON SYDNEY AUCKLAND TORONTO

Preface

Although A-level syllabuses in Biology have undergone relatively few changes in the last decade, methods of examining have changed. Essays and longer structured questions, which at one time were the major component of all examination papers, have been replaced by a more varied type of question, including multiple choice, short structured questions, the handling of data, the construction of graphs and the solution of problems.

This book contains a collection of original questions of all types, similar to those set in GCE A-level examinations, and grouped according to the major sub-divisions of the majority of A-level syllabuses. Apart from enabling teachers to undertake diagnostic testing of their classes, the book should provide students with an aid to study, encouraging them to learn intelligently and to read texts with particular questions in mind.

I hope that, as a result of working through the problems, students will be stretched to their full intellectual capacity and thereby gain a clearer, deeper understanding of material that forms the core of most A-level syllabuses in Biology.

Permission to reproduce and to make use of data, some in a modified form, is gratefully acknowledged:

Blackwell Scientific Publications Ltd, and Mr B Walker (Q. 21c, Ch. 11) The problem of the migration of the european eel (*Anguilla anguilla*). Berry, L., Brookes, D. and Walker, B. (1972) *Science Progress* **60** (240), 465–485.
The Association for Science Education and Mr I. A. Dodd (Fig. 61). 'Biology Notes', *School Science Review* (1964 and 1966).
Hodder and Stoughton Ltd (Q. 21 and Q. 25c and d, Ch. 9). R. Dajoz (1977) *Introduction to Ecology*.
Fig. 11.7 is based on the system of classification devised by Whittaker, *Science*, **163**, 150–160 (1969).

In addition, my thanks are due to Mrs Marta Moore and to many of my students, both past and present, who helped either to assemble data or to work through the questions, providing useful comments. Any errors which have been overlooked will, I hope, be treated with tolerance and reported to me for future amendment.

P. W. F.

Also by P. W. Freeland
Problems in Practical Advanced Level Biology

Contents

British Library Cataloguing in Publication Data

Freeland, P. W.
 Problems in theoretical Advanced Level biology.
 1. Biology—Examinations, questions, etc.
 I. Title
 574'.076   QH316

 ISBN 0-340-333564-5

First printed 1985

Typeset in 10/11pt Univers (Monophoto) by Macmillan India Ltd, Bangalore

Printed in Great Britain for
Hodder and Stoughton Educational,
a division of Hodder and Stoughton Ltd,
Mill Road, Dunton Green, Sevenoaks, Kent TN13 2YD
at the University Press, Oxford.

Advice to Students

Each year around 40 000 students are entered for A-level examinations in Biology, and of these rather more than 65 per cent are generally successful in reaching the required standard. Biology, like many other academic subjects, is one in which there is no substitute for hard work, sustained throughout the period of study; time has to be spent in reading texts and in answering questions that help to promote a full understanding of factual material.

Whilst teachers can do a great deal to help students develop an understanding of biological facts and principles and an appreciation of their significance, the student must show sufficient self-motivation and interest to set aside a fixed amount of time each week for private study. Moreover, that period of private study needs to be sub-divided into a time for learning, revision and practice, modified in the light of academic progress. Those who mark examination papers are confronted repeatedly by similar, if not identical, failings, many of which could be avoided if students had, from the outset, a clearer understanding of what was required of them. Whilst it would be impossible to state all of these requirements, some of the more important points are listed below.

The syllabus

Each Examining Board publishes a syllabus, which lists topics to be studied and indicates particular aspects of these topics that require special consideration. Make sure that you possess your own copy of the syllabus, that you read it, and that you mark off topics as they are covered. If you are unfamiliar with your syllabus, certain topics may be omitted, leaving you at a disadvantage.

Textbooks

In Biology, as in all the sciences, new knowledge is continually becoming available as a result of research. Within those topics specified by the syllabus, your knowledge is expected to be both wide-ranging and up-to-date. Therefore, no single textbook can ever meet all your needs, nor can it answer all the questions that examiners are likely to set. Wherever possible, work from two or more general textbooks, reading through different accounts of the same topic, especially if they are written from different viewpoints. Take particular note of keywords, generally printed in heavy type, and make use of these words in your own writings. At the same time, make use of the school library or public library service, referring occasionally to more advanced texts, such as those dealing specifically with plant physiology, animal physiology, cytology, genetics or evolution. Having completed your study of a particular area of the syllabus, look at past examination papers, or at one of the chapters in this book, to make certain that you have grasped the main points and possess a sufficient knowledge to provide wide-ranging, accurate, concise or full answers that you can shape to provide answers to particular questions.

Essays

The primary requirement in an essay is that factual material should be organised into a number of paragraphs, each dealing with a particular aspect of the subject. Hence, essays on biological topics might, in general, be organised along the following lines.

(i) Introduction: definitions, historical aspects.
(ii) Paragraph A: occurrence, nature, structure, distribution.
(iii) Paragraph B: functions, mechanisms.
(iv) Paragraph C: biological significance, evolutionary significance, commercial significance.
(v) Conclusions: a summary of findings, indications for further research.

Secondly, essays need to be full, detailed accounts of the topics under discussion. Whilst a lengthy essay, covering 4–6 sides of paper, is not necessarily a good essay, a short essay, covering only 1–2 sides of paper, is invariably a poor one. Aim, therefore, to cover 3–4 sides of A4 paper in each half-hour period. Don't, however, be misled into thinking that an essay on a biological topic should consist of continuous prose, and that alone. Good essays are always accompanied by diagrams, graphs, flow-charts, tables, chemical equations etc. If you omit these essential ingredients, you are unlikely to be awarded high marks for your efforts.

Long structured questions

Read these questions carefully and answer the questions that are set. Do not resort to setting and answering your own variants. Invitations to 'describe' particular objects or structures, require both a written account and illustrations, often in the form of a diagram, flow-chart or graph. Discussions, on the other hand, generally require the presentation of alternative viewpoints, an explanation of possible alternative mechanisms, or an overview of a topic that requires clarification through further research.

Pay particular attention to the marks awarded for each part of these questions. Keep one eye on the clock and do not spend a disproportionate amount of time on parts of questions that carry only a few marks. On the other hand, bear in mind that definitions need to be full. Therefore, if 3–4 marks are awarded for a definition, aim to write 3–4 sentences, covering perhaps as much as one-third of a page. Remember that modern Biology is a technical subject, with its own vocabulary of technical terms and phrases. Make appropriate use of these terms and phrases at every opportunity. Pepper your writing with the vocabulary of science. Additionally, remember to make appropriate use of diagrams, graphs and tables, never relying on the written word alone to communicate the points you wish to make.

Short structured questions

Short structured questions require a detailed knowledge of particular areas of the syllabus, as well as a logical approach, clear thought, and the ability to interpret data and solve problems. Additionally, and perhaps most important of all, short structured questions require skill in the use of words. Again, it is important to answer the questions that are set, confining each answer to the space provided. This is an art which cannot be perfected without practice. When preparing for this type of question, pay particular attention to those drawings or diagrams that appear in all the general textbooks, noting the functions played by each of the labelled items.

Weaker candidates may find difficulties, within the time limit available, in answering all the short structured questions set on the paper. Once again, the remedy for this situation lies in further practice, working always against the clock.

Multiple choice questions

This type of question is set by some, but not by all, of the Examining Boards. Multiple choice questions are designed to test the breadth and depth of your knowledge, as well as penalising those who have not covered all areas of the syllabus. Many students approach these questions in an over-confident frame of mind, answering them hurriedly, without giving adequate consideration to all the alternative options. Although knowledge can only be acquired as a result of learning, practice in answering this type of question often has a beneficial effect on performance.

Data-handling questions

Information presented in the form of data may require any one of a number of treatments. If a discussion is required, try to look beyond the data to the biological processes upon which the data are based. Give an outline account of these processes, view the data against this background, explain the significance of the data, and draw conclusions, or make recommendations, depending on the nature of the question. If you are required to present the data in a visual form, consider if a pie chart, histogram or graph would be most appropriate. Use a pie chart only when data is composed of figures that add up to 1.0 or 100%, with each of the figures forming a component fraction of the whole unit. Use a histogram for measurements of height, length etc, where data relates to particular classes or categories, such as height within the age range 0–5, 6–10. Generally, plot data as a graph, but bear in mind that certain conventions must be followed if high marks are to be attained.

(i) Draw a large graph, to fill the graph paper or available space.
(ii) Plot time, or regular units, along the x (horizontal) axis. Plot the dependent variable along the y (vertical) axis.
(iii) Plot specific points on your graph as dots, joining them by a series of straight lines. Do not attempt to draw 'best-fit' curves.
(iv) Name the units along the x and y axes. Provide a title.
(v) In graphs that require more than one vertical scale, plot one scale to the left, and the other to the right, of the horizontal scale. Use different coloured ball-point pens to relate a particular graph to a particular scale.

Experiments

Accounts of experiments, carried out either in the laboratory, or in the field, should generally be presented in accordance with the following format.

(i) Title.
(ii) Objectives: a statement of aims.
(iii) Materials: a list of requirements.
(iv) Method: a logical, ordered statement of procedures.
(v) Results: the outcome of the experiment, presented in the form of tables, graphs etc.
(vi) Discussion: including difficulties in the use of apparatus, limitations of the techniques employed, conclusions, and proposals for further research.

If an examiner asks for experimental techniques that could be applied to particular biological materials, such as the epidermis of a leaf or the alimentary canal of a mammal, the method described must, of course, be applicable to that material, and capable of yielding the information that is required.

1 Macromolecules

Carbohydrates, fats, proteins, nucleic acids, enzymes, vitamins, animal hormones, phytohormones

Multiple Choice Questions

1 Which of the following is a hexose sugar?
(a) glycogen
(b) lactose
(c) xylose
(d) ribulose
(e) fructose.

2 Which of the following could *not* be classified as a lipid?
(a) fats and oils
(b) photosynthetic pigments
(c) steroids
(d) waxes
(e) anthocyanin pigments.

3 Trypsin, insulin, actin, myosin and haemoglobin are all classified as
(a) carbohydrates
(b) fats
(c) nucleic acids
(d) proteins
(e) enzymes.

4 Semiconservative replication of DNA was first demonstrated by
(a) Watson and Crick
(b) Herschey and Chase
(c) Meselson and Stahl
(d) Beadle and Tatum
(e) Jacob and Monod.

5 Which of the following bases occurs in DNA but not in RNA?
(a) thymine
(b) uracil
(c) guanine
(d) cytosine
(e) adenine.

6 The synthesis of mRNA on a DNA template is called
(a) translocation
(b) transcription
(c) transpiration
(d) translation
(e) transportation.

7 DNA, RNA and ATP are composed of
(a) pentose sugars
(b) nucleotides
(c) purines
(d) pyrimidines
(e) nucleic acids.

8 The anticodon of a tRNA molecule consists of the base sequence adenine(A) – adenine (A) – uracil(U). The appropriate codon of a mRNA molecule will therefore be
(a) U–U–A
(b) A–U–U
(c) A–A–U
(d) U–A–A
(e) T–T–G.

9 Which of the following statements is *not* universally applicable to enzymes?
(a) They generally work very rapidly.
(b) They can catalyse a reaction in both directions.
(c) They are not used up during a reaction.
(d) They will bind one substrate only.
(e) They are proteins.

10 Vitamin A is an essential dietary factor for the formation of
(a) trypsin
(b) biotin
(c) rhodopsin
(d) biliverdin
(e) haemoglobin.

11 Vitamin B_1 is a co-enzyme essential for the oxidation of
(a) pyruvic acid
(b) nucleic acid
(c) ATP
(d) vitamin C
(e) proteins.

12 One of the following is both a vitamin and a hormone. Is it
(a) ascorbic acid (vitamin C)
(b) calciferol (vitamin D)
(c) thiamin (vitamin B_1)
(d) riboflavin (vitamin B_2)
(e) α-tocopherol (vitamin E)?

13 The primary effect of gibberellin is on
(a) mitosis
(b) meiosis
(c) cell enlargement
(d) flowering
(e) root growth.

Short Structured Questions

14 Name a carbohydrate that is
(a) stored in mammalian liver and skeletal muscles
(b) stored in the cells of vascular plants
(c) the major component of primary cell walls
(d) translocated in the phloem
(e) an intermediate product in the conversion of starch to glucose and fructose
(f) used as respirable substrate by milk-fermenting bacteria
(g) transported in the blood stream
(h) an isomer of glucose
(i) a component of RNA
(j) a component of DNA.

15 Fig. 1 shows the structural formula of a polypeptide molecule.
(a) On a copy of the diagram, indicate the position of the following regions in the molecule
A: an amino group
B: a carboxyl group
C: a peptide bond
D: a ring of carbon atoms
E: a hydroxyl group.
(b) Show the structural formula of the polypeptide if a second molecule of glycine were to be added to it to form a dipeptide.
(c) What chemical terms are used to describe
(i) the addition of further amino acids to a peptide
(ii) the removal of amino acids from a peptide?

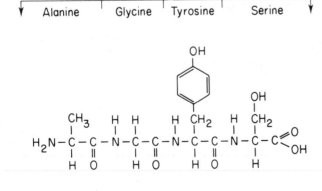

Fig. 1   A polypeptide

16 (a) Name four proteins, other than enzymes, which play different roles in the body of a mammal. State the role of each protein you have named.
(b) What types of bonding occur in protein molecules?
(c) Outline a simple test that will demonstrate that a substance contains protein.

17 In the sequencing of amino acids in a polypeptide, the polypeptide is hydrolysed by enzymes and smaller fragments identified by paper chromatography. The following peptide fragments, identified by the initial letters of their amino acids, were isolated from the random hydrolysis of a peptide:
Glu–ala
Ser–tyr
Ser–tyr–leu
Arg–glu
His–lys–COOH
Asp–arg–glu–ilu–his
Tyr–leu–asp–arg.
What short amino acid sequence is consistent with these fragments?

18 It is known that triplets of bases in DNA molecules code for specific amino acids. Some of these base sequences and the amino acids for which they code are given in Table 1.
A cell synthesises a polypeptide consisting of the following amino acid sequence:
Ala–ala–lys–leu–lys–arg.
(a) What is the sequence of bases in the DNA template?
(b) What is the sequence of bases in the mRNA molecule formed on the DNA template?
(c) What is the sequence of bases in the complimentary region of the DNA molecule that played no part in the synthesis?
(d) If the DNA molecule was found to contain 32% adenine, what percentage of guanine would you expect to find in the (i) DNA molecule and (ii) mRNA molecule?

Table 1

DNA triplet	Amino acid
CGA	Alanine (Ala)
GCA	Arginine (Arg)
AAT	Leucine (Leu)
TTT	Lysine (Lys)

19 Fig. 2 illustrates the Fischer lock and key hypothesis of enzyme activity.
(a) Give a short account of the Fischer theory.
(b) What name is given to molecule A, which must attach to the enzyme before it can function as a catalyst? Name two vitamin-derived compounds that may act in this capacity.

(c) Molecules B and C are enzyme substrates. At which region of the enzyme do they become attached?

(d) Molecule D is an enzyme activator. Name two mineral ions that may act in this role and the enzymes with which they are associated.

(e) Molecule E is a non-competitive inhibitor, bound to the allosteric site. How is such a molecule believed to exert its influence? Name a non-competitive inhibitor used as an insecticide and the enzyme that it inactivates.

(f) Molecule F is a competitive inhibitor. How does it affect the rate of an enzyme-catalysed reaction? Give the name of a compound known to act as a competitive inhibitor of a named reaction.

(g) What general name might be applied to molecule G?

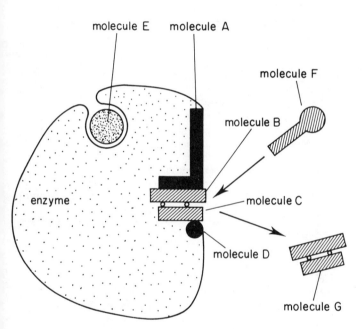

Fig. 2 Fischer's lock and key model of enzyme activity

20 Table 2 lists a number of enzyme substrates. Complete the table by naming an appropriate enzyme and the products of the reaction.

Table 2

Substrate	Enzyme	Reaction products
(a) Acetylcholine		
(b) Fibrinogen		
(c) Hydrogen peroxide		
(d) Polypeptides		
(e) Urea		

21 In an investigation of phosphatase activity in the testis of a rat, unit masses of the testis were macerated in a mortar and then mixed with 10 cm^3 of a 0.1 % solution of phenolphthalein diphosphate, an artificial enzyme substrate that yields a coloured product when degraded by the enzyme. Investigations were made into the effects of temperature and pH on enzyme activity. Records were made of the time taken for the contents of each tube to reach the same colour intensity as that of a standard solution. The results obtained are given in Table 3.

(a) Plot these results on the same graph to show the relationship that exists between the rate of the reaction, temperature and pH,

(b) What conclusions do you draw from the results?

(c) How could you improve the accuracy of the graphs which have been drawn from the data?

Table 3

Environmental condition		Time taken for completion of the reaction (h)
A. At optimum pH Temperature (°C)	0	5.3
	10	2.4
	20	0.9
	30	0.4
	40	0.3
	50	0.9
	60	Not completed
	70	Not completed
B. At optimum temperature pH	3.0	0.9
	4.0	0.5
	5.0	0.7
	6.0	0.8
	7.0	1.6
	8.0	0.6
	9.0	0.3
	10.0	0.6
	11.0	0.8
	12.0	0.9

22 The incomplete graphs, A to D in Fig. 3, refer to different aspects of enzyme activity.

(a) Give a concise definition of an enzyme.

(b) The enzyme pepsin, produced in the stomach, catalyses the conversion of proteins into proteoses and peptones. From your knowledge of the enzyme, complete the graphs A to D relating pepsin activity to pH, temperature, enzyme concentration and substrate concentration.

(c) What assumption is made in D?

(d) How would your graphs have differed if trypsin, rather than pepsin, had been used?

(e) What do you understand by the Q_{10} of an enzyme-controlled reaction?

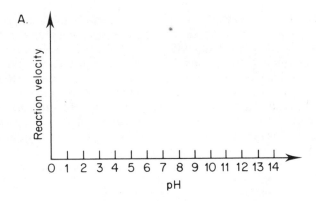

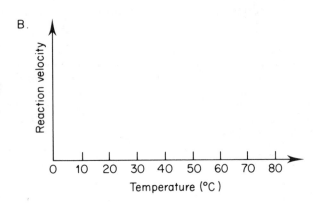

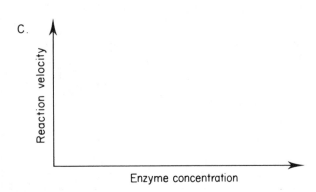

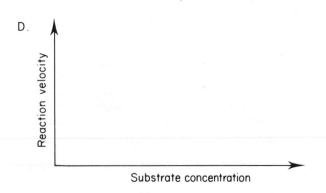

Fig. 3 Aspects of enzyme activity

23 Some characteristics of vitamins B$_1$, B$_6$, C and D are listed below. Re-list these characteristics, by their initial letter only, against the appropriate vitamin.
(a) It is also known as ascorbic acid.
(b) Deficiency of this vitamin causes beri-beri.
(c) It is a hormone as well as a vitamin.
(d) It is unstable in hot alkaline solutions.
(e) It is synthesised by most mammals, but not by primates.
(f) It is a powerful reducing agent.
(g) This vitamin is fat-soluble.
(h) It is not derived entirely from the human diet.
(i) It is a co-enzyme in the formation of haemo-globin and is also required for amino acid metabolism.
(j) Potatoes are an important dietary source of this vitamin for people living in the British Isles.

24 Table 4 lists some animal hormones, their source and their effects. Complete the table by filling in the blank spaces.

Table 4

	Hormone	Source of hormone	Effects of hormone
(a)	Insulin		
(b)		Adrenal medulla	
(c)			Anti-inflammatory and anti-allergic
(d)	Antidiuretic hormone (ADH)		
(e)			Promotes growth of graafian follicles
(f)		Thyroid gland	
(g)	Oxytocin		

25 Some characteristics of the phytohormones gibberellin, auxin and ethene (ethylene) are listed below.
(a) It is used to prevent petal-drop and fruit-drop in apple orchards.
(b) It substitutes for light in the germination of light-sensitive seeds of lettuce, tobacco and mistletoe.
(c) It stimulates cell division in association with cytokinins.
(d) It was originally discovered by Japanese physiologists working with diseased rice plants.
(e) It prevents straightening of the plumule in pea seedlings.
(f) It stimulates swelling of the receptable in strawberries and apples.

(g) It promotes the formation of staminate (male) flowers in hop and cucumber.

Re-list these characteristics, by their initial letter only, against the appropriate phytohormone.

26 Some structural formulae of compounds that occur in living organisms are given below:

(a) CH_2OH
  |
  $CHOH$
  |
  CH_2OH

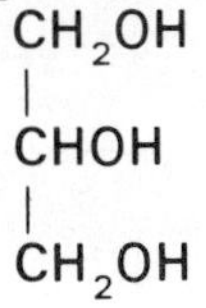

(b) $COOH$
  |
  $C=O$
  |
  CH_2
  |
  CH_2
  |
  $COOH$

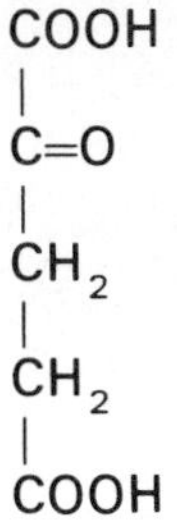

(c) $COOH$
  |
  $H-C-NH_2$
  |
  CH_2
  |
  $COOH$

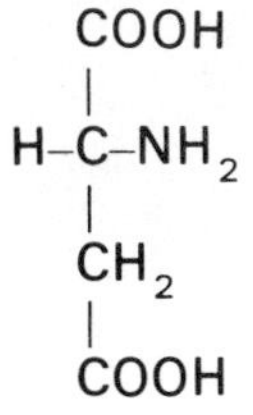

(d) CH_2-COOH

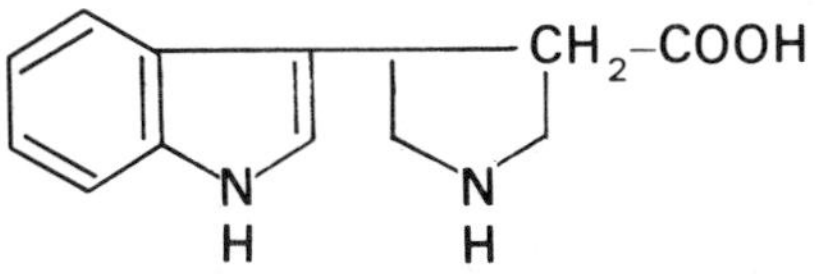

(e) HO — — $CH(OH)CH_2NH . CH_3$
  HO

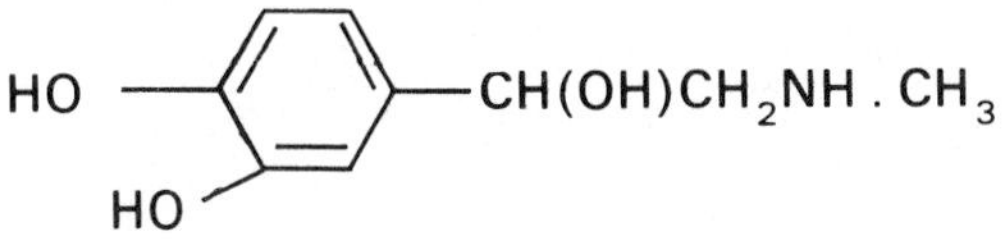

(f) $CH(CH_3) . CH_2 . CH_2 . CH_2 . CH(CH_3)_2$

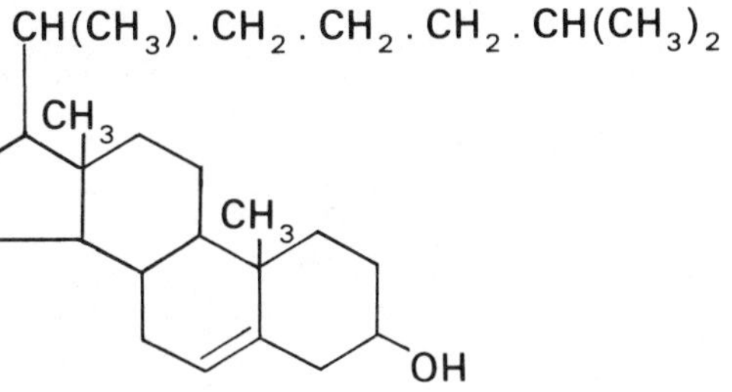

(g) $C_{15}H_{31}COOH$

(h) $CH_2O . CO . C_{17}H_{35}$
  |
  $CHO . CO . C_{17}H_{35}$
  |
  $CH_2O . CO . C_{17}H_{35}$

(i)

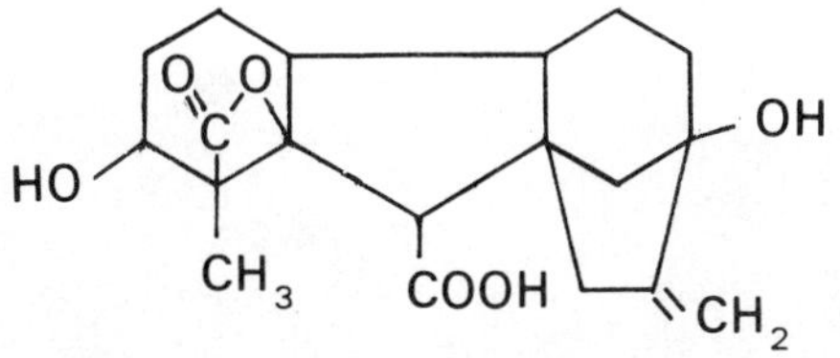

(j)

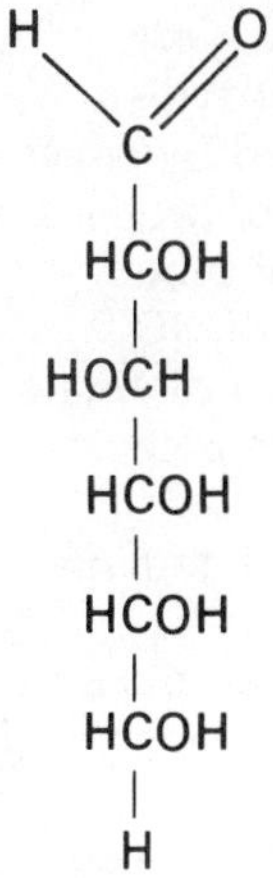

$H-C$ (with $=O$)
  |
  $HCOH$
  |
  $HOCH$
  |
  $HCOH$
  |
  $HCOH$
  |
  $HCOH$
  |
  H

Which of these formulae represents the following compounds?
  (i)   a saturated fat
  (ii) glucose
  (iii) cholesterol
  (iv) glycerol
  (v)   adrenalin
  (vi) gibberellin
  (vii) an amino acid
  (viii) a keto acid
  (ix) an unsaturated fatty acid
  (x)   indoleacetic acid (IAA).

27 Construct a diagram to illustrate the classification of named carbohydrates into groups and sub-groups.
  Describe how the molecular structure of carbohydrates is related to their functional role in cells as
  (a) an energy source
  (b) a food reserve
  (c) a skeletal component.

28 What is a lipid?
Give a description of the variety, distribution and functions of those lipids that are components of membranes or are secreted by cells.

29 What functions do proteins perform in living cells? Show, with reference to a named protein, how its function is dependent upon its shape.
  Describe the stages by which proteins are synthesised in cells by sequential assembly of amino acids.

30 Explain the meaning of the following terms which are applicable to a carbohydrate, fat or protein: reducing and non-reducing sugar; polysaccharide; polymerisation; isomerism; emulsification; neutral fat; condensation and hydrolysis; denaturation.

31 By reference to enzymes, explain the following terms and concepts: protein conformation; activation energy; active site; lock and key hypothesis; specificity; induction; optimal activity; co-enzyme; competitive and non-competitive inhibition; denaturation.

32   Define the term *enzyme*.
List the properties of enzymes.
Describe the role played by enzymes in
    (a) the digestion of proteins in the gut of a mammal
    (b) the clotting of blood
    (c) the transport of CO_2
    (d) the contraction of striated muscle
    (e) germination of a cereal grain.

33   Write an essay on vitamins.

34   Distinguish between a *hormone* and a *pheromone*.
State the general importance of hormones and phero-
mones in mammals.
Show, with reference to named examples,
    (a) how hormones effect co-ordination between
        different organs in the body of a mammal
    (b) how pheromones effect co-ordination between
        different members of a population of mammals.

35   What are phytohormones?
Describe some of the effects of phytohormones on
      (a) the cells and
      (b) the organs of a named vascular plant.
List the practical uses of phytohormones in horticulture
and agriculture.

36   Illustrate and discuss the view that the macromole-
cules found within cells serve either as structural components or are concerned with the storage and transmission of information.

37   Give an account of practical techniques that would
enable you to
    (a) separate and identify amino acids produced by
        the complete hydrolysis of a peptide
    (b) separate cellulose from a powder containing
        cellulose, starch and glucose
    (c) demonstrate the effects of substrate concentra-
        tions on the rate of reaction of a named enzyme
    (d) measure the concentration of vitamin C in
        orange juice
    (e) bioassay the concentration of auxin in
        the shoot tip of a maize seedling.

2 Cells, Organelles and Membranes

Prokaryotic cells, eukaryotic cells, cell organelles, mitosis, meiosis

Multiple Choice Questions

1 Prokaryotic cells include
   (a) bacteria
   (b) blue-green algae
   (c) bacteria and blue-green algae
   (d) fungi
   (e) protozoa.

2 Which of the following would *always* contain some secretory cells?
   (a) lenticels
   (b) hydathodes
   (c) xylem
   (d) phloem
   (e) epidermis.

3 In which of the following tissues would you find cells that generally performed a protective function?
   (a) vascular cambium
   (d) xylem
   (c) epidermis
   (d) phloem
   (e) cortex.

4 Suberin impregnates the cell wall in
   (a) xylem
   (b) cork
   (c) parenchyma
   (d) phloem
   (e) collenchyma.

5 The principal respiratory, transpiratory and photosynthetic cells of a herbaceous plant are found in the
   (a) mesophyll of a leaf
   (b) cortex of a stem
   (c) cortex of a root
   (d) root hair region
   (e) vascular bundle.

6 The term sclerenchyma may be applied to
   (a) fibres
   (b) sclereids

   (c) fibres and sclereids
   (d) fibres, sclereids and xylem vessels
   (e) fibres, sclereids, xylem vessels and medullary rays.

7 A correct identification of collenchyma is dependent upon a cell possessing
   (a) chloroplasts
   (b) a secondarily thickened wall
   (c) a primary wall, thicker in some regions than in others
   (d) a primary wall of even thickness
   (e) a lignified cell wall.

8 In the leaves of herbaceous plants, chloroplasts would generally occur in the
   (a) mesophyll
   (b) epidermis
   (c) bundle sheath
   (d) vascular tissue
   (e) palisade parenchyma, spongy parenchyma and guard cells.

9 The primary wall of parenchyma cells consists of
   (a) cellulose
   (b) cellulose and lignin
   (c) cellulose and hemicellulose
   (d) cellulose, hemicellulose and pectic acid
   (e) cellulose, hemicellulose, pectic acid and cholesterol.

10 Bordered pits occur between adjacent cells in
   (a) fibres
   (b) xylem vessels
   (c) fibres and xylem vessels
   (d) parenchyma
   (e) collenchyma.

11 Which of these organelles makes the most direct contribution to the formation of cellulose cell walls in plants?
   (a) lysosomes
   (b) ribosomes
   (c) plastids
   (d) Golgi bodies
   (e) mitochondria.

12 A secretory and conducting function is performed by
   (a) osteocytes
   (b) erythrocytes
   (c) neurons
   (d) chondrocytes
   (e) leucocytes.

13 Complex pores are an important structural feature of the
- (a) cell surface membrane
- (b) nuclear envelope
- (c) endoplasmic reticulum
- (d) Golgi body
- (e) chloroplast.

14 The nucleic acid–protein complex of a chromosome is called
- (a) DNA
- (b) histone
- (c) chromatin
- (d) nucleolin
- (e) nucleolus.

15 In which of the following would you expect to find the largest number of mitochondria per cell?
- (a) flight muscles of an insect
- (b) root hairs of a vascular plant
- (c) cotyledons of a seed
- (d) testis of a mammal
- (e) liver of a mammal.

16 Microtubules are a component of
- (a) cilia
- (b) ribosomes
- (c) mitochondria
- (d) nuclei
- (e) Golgi bodies.

17 A chromosome attaches to the spindle by means of its
- (a) chromatid
- (b) centrosome
- (c) centriole
- (d) centromere
- (e) chromomere.

18 In which of the following does the ornithine cycle occur?
- (a) ribosomes
- (b) mitochondria
- (c) lysosomes
- (d) chloroplasts
- (e) nucleus.

19 In which of these phases of mitosis would the amount of DNA per cell nucleus be one-half of that in late interphase?
- (a) prophase
- (b) metaphase
- (c) anaphase
- (d) telophase
- (e) anaphase and telophase.

20 A typical cell cycle involves growth (G_1 and G_2), synthesis of DNA (S) and cell division (mitosis and meiosis). In which of the following sequences do these events generally occur?
- (a) $G_1 : G_2 : S$: mitosis
- (b) $S : G_1 : G_2$: mitosis
- (c) $G_1 : S : G_2$: mitosis
- (d) $S : G_1 : G_2$: meiosis
- (e) $S : G_2 : G_1$: mitosis.

21 Where, in a typical herbaceous plant, would you expect to find cells which showed the following features or performed the following functions?
- (a) store water
- (b) store starch
- (c) absorb water and mineral salts
- (d) possess walls thickened by deposits of lignin
- (e) effect bidirectional translocation of solutes
- (f) respond to gravity
- (g) show diurnal variations in water content
- (h) effect a unidirectional translocation of water and mineral salts
- (i) perform photosynthesis
- (j) exhibit stages in mitosis
- (k) show seasonal meiosis.

22 In which tissues or organs of a mammal would you find the following cells?
- (a) osteocytes
- (b) chondrocytes
- (c) neurons
- (d) Purkinje fibres
- (e) Pacinian corpuscles
- (f) oxyntic cells
- (g) glial cells
- (h) neutrophils
- (i) ciliated epithelium
- (j) squamous epithelium
- (k) spermatocytes
- (l) rods
- (m) Schwann cells.

23 With which cell organelles, or membrane systems, do you associate each of the following processes?
- (a) photolysis of water
- (b) synthesis of polypeptides
- (c) oxidative phosphorylation
- (d) synthesis of ribosomal RNA
- (e) enzyme secretion
- (f) organisation of the mitotic spindle
- (g) production of CO_2
- (h) osmoregulation
- (i) digestion of foreign inclusions
- (j) communication between adjacent cells
- (k) water storage
- (l) cell locomotion.

24 Which of the following statements relate to mitosis and which to meiosis?
- (a) $N \rightarrow N + N$
- (b) $2N \rightarrow 2N + 2N$
- (c) $2N \rightarrow N + N + N + N$
- (d) typically observed in meristematic cells
- (e) a reduction division

(f) comprises two successive divisions of a cell
(g) prophase involves the formation of bivalents
(h) crossing-over effects an exchange of genes
(i) prophase is complex, involving several stages
(j) the only type of division that occurs in a game-tophyte plant.

25 Fig. 4 shows a particular stage in mitosis in which a pair of homologous chromosomes is involved.
  (a) Identify the structures A to F in the diagram.
  (b) Name the stage in mitosis that is illustrated. What stages (i) have preceded it, and (ii) will follow it?
  (c) Make drawings to show the appearance of these two chromosomes in the corresponding stages of meiosis. Write a brief explanation of your drawings.

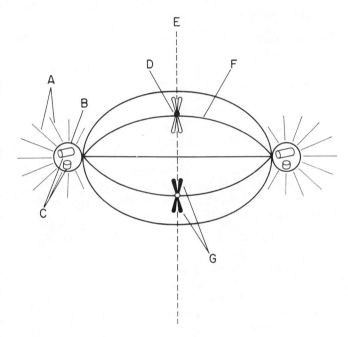

Fig. 4 The appearance of a stage in mitosis

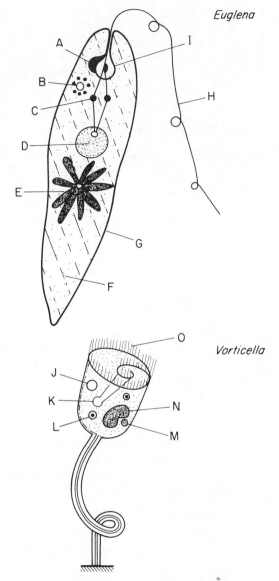

Fig. 5 Diagrams of *Euglena* and *Vorticella*

26 The organisms shown in Fig. 5 are *Euglena* and *Vorticella*.
  (a) Identify the structural features A to O in the diagram.
  (b) Which of these organisms would be
      (i) found in ponds
      (ii) classified as a ciliate
      (iii) motile
      (iv) heterotrophic
      (v) most closely related to *Trypanosoma*
      (vi) most closely related to *Paramecium*
      (vii) regarded as showing features typical of both plants and animals
      (viii) capable of reproduction by longitudinal fission
      (ix) capable of exhibiting a metachronal rhythm
      (x) employed in the disposal of sewage?

27 Fig. 6 illustrates two cell organelles.
  (a) Identify the organelles X and Y and name a type of cell in which each might occur.
  (b) Identify the structural features A to J in the diagram.
  (c) Indicate as precisely as possible the location of the following compounds within the organelles.
      (i) enzymes of the TCA cycle (Krebs' cycle)
      (ii) enzymes of the Calvin cycle
      (iii) enzymes associated with photolysis
      (iv) enzymes of oxidative phosphorylation
      (v) cytochromes.
  (d) Which of the following statements is true for organelles of type X and organelles of type Y.
      (i) They release oxygen.
      (ii) They release carbon dioxide.
      (iii) They occur in the largest numbers in the body of a green plant.

(iv) They are generally of the larger size and mass.
(v) They are absent from bacteria.
(vi) They synthesise ATP and NADPH.
(vii) They contain DNA and ribosomes.
(viii) They are associated with the Calvin cycle.
(ix) They are associated with the ornithine cycle.
(x) Their inner membrane is folded into layers.

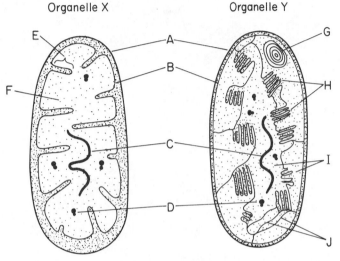

Fig. 6   Diagrammatic representation of two cell organelles

28   Fig. 7 illustrates two models of a cell membrane.
(a) Name model X and model Y.
(b) Identify the component molecules, labelled A to D, in the two models.
(c) Explain why biologists have generally adopted model B in preference to model A.

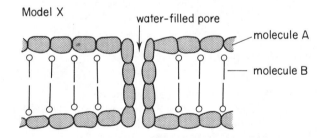

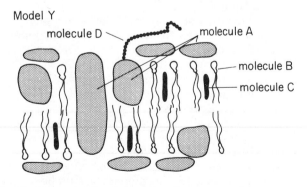

Fig. 7   Two models of a cell membrane

Long Structured Questions

29   Describe, in detail, how you would prepare material for examination by
(a) a light microscope
(b) an electron microscope.
How is a magnified image produced by each of these instruments? What are the limitations of light microscopes and electron microscopes?

30   Evaluate the contributions made by each of the following to our understanding of cell structure and function.
(a) light microscopy   (e) chromatography
(b) electron microscopy   (f) electrophoresis
(c) X-ray crystallography (g) radioactive isotopes.
(d) centrifugation

31   By means of large, labelled drawings illustrate the ultrastructure of
(a) a generalised cell
(b) a generalised plant cell and
(c) a generalised animal cell.
How would specialised cells such as (i) neurons, (ii) striated muscle fibres and (iii) sieve elements differ from the generalised cells you have drawn?

32   Show, by reference to named free-living unicells, how relatively simple organisms may possess complexity in structure, diversity in modes of nutrition and variety in modes of reproduction.

33   Describe the structural and functional differences that exist between
(a) cell membrane and cell wall
(b) nucleus and nucleolus
(c) microtubule and microfilament
(d) ribosome and lysosome
(e) cilia and flagella
(f) chromosome and centrosome
(g) endoplasmic reticulum and Golgi body.

34   Account for ATP synthesis in chloroplasts and mitochondria.

35   Indicate by reference to specific examples how materials enter or leave cells by
(a) diffusion   (c) phagocytosis and pinocytosis
(b) osmosis   (d) active transport.

36   How would you demonstrate the following experimentally?
(a) A cell extract containing mitochondria can effect release of CO_2 from a substrate such as succinate.
(b) Living plant material contains active mitochondria.
(c) A cell extract containing chloroplasts is capable of effecting reduction of an artificial hydrogen acceptor such as DCPIP.

3 Water, Solutes and Transport Systems

Water, solutes, osmoregulation, xylem, phloem, transpiration, stomata, blood system

Multiple Choice Questions

1 In the terminology currently used by plant physiologists, the term *solute potential* is equivalent in meaning (but opposite in sign) to
(a) osmosis
(b) osmotic pressure
(c) osmotic potential
(d) wall pressure
(e) turgor pressure.

2 A plant cell that had lost a large amount of water from its cell vacuole would most probably be in a state of
(a) plasmolysis
(b) turgor
(c) endosmosis
(d) osmosis
(e) diffusion pressure deficit.

3 Calcium is an essential nutrient for green plants as it
(a) forms salts that bind cellulose molecules in cell walls
(b) is required for active transport
(c) forms part of the cytochrome molecule
(d) is a component of DNA
(e) is a co-enzyme for rennin.

4 Green vascular plants normally utilise inorganic nitrogen in the form of
(a) NO_2^-
(b) NO_3^-
(c) NH_4^+
(d) NO_3^- and NH_4^+
(e) amino acids.

5 Iron is an essential component of the human diet because it is a component of
(a) myosin
(b) haemoglobin
(c) nucleic acids
(d) adrenalin
(d) ferredoxin.

6 Xylem is composed of
(a) fibres
(b) vessels
(c) fibres, vessels and tracheids
(d) fibres, vessels and medullary ray parenchyma
(e) fibres, vessels, tracheids and medullary ray parenchyma.

7 Which of the following is *not* translocated principally through the phloem of a vascular plant?
(a) sucrose
(b) auxin
(c) K^+ ions
(d) water
(e) amino acids.

8 Included amongst phagocytic cells of the blood are
(a) monocytes
(b) neutrophils
(c) monocytes and neutrophils
(d) monocytes, neutrophils and lymphocytes
(e) monocytes, neutrophils, lymphocytes and platelets.

9 Antibodies are a product of
(a) granulocytes
(b) lymphocytes
(c) neutrophils
(d) monocytes
(e) erythrocytes.

10 In which of the following forms is most CO_2 carried in the blood stream?
(a) CO_2 molecules in solution
(b) carbonic acid
(c) bicarbonate
(d) carbamino protein
(e) carboxyhaemoglobin.

11 Three blood vessels A, B and C have diameters in the ratio $0.85 : 1.0 : 1.2$. The resistance to blood flow in these vessels will be in the ratio
(a) $0.85 : 1.0 : 1.2$
(b) $0.72 : 1.0 : 1.4$
(c) $0.6 : 1.0 : 1.7$
(d) $2.0 : 1.0 : 0.5$
(e) $500 : 1.0 : 0.005$.

Short Structured Questions

12 Fig. 8 is a diagram of two adjacent plant cells.
    (a) Identify the structures A to F in the diagram.
    (b) In passing from one cell to another, water may pass along one of two different routes. What are these routes?
    (c) Cells X and Y are just turgid. Describe, using appropriate terms, the changes that would occur if the cells were placed into (i) molar glucose solution and (ii) distilled water.

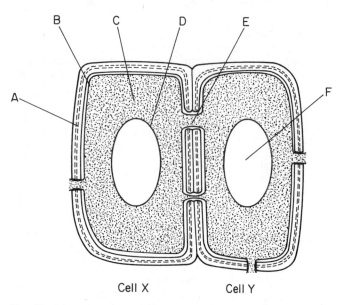

Fig. 8 Diagram of two adjacent plant cells

    (d) Plant physiologists prefer to describe the water relationships of plant cells in terms of the general equation

$$\psi \quad = \quad \psi_S \quad + \quad \psi_P$$
Water potential   Solute potential   Pressure potential

Plot the figures given in Table 5 as a graph.
    (e) Calculate values for ψ and plot these on your graph.
    (f) Define the term *water potential*.
    (g) What is the water potential of pure water?
    (h) What effect does the addition of solutes have on the water potential?
    (i) Why is the water potential of a fully turgid cell never equal to 0?
    (j) State one advantage of the use of water potential terminology.
    (k) Cells X and Y are adjacent to one another in the same plant. In cell X $\psi_p = +0.5$ MPa and $\psi_s = -1.2$ MPa, whereas in cell B $\psi_p = +0.3$ MPa and $\psi_s = -0.6$ MPa.
    (i) What is the water potential in each of these cells?
    (ii) In which direction will water move and with what force?

Table 5

Relative cell volume	Potential ψ_p	(MPa) ψ_s
0.8	−0.3	−2.8
0.9	−0.2	−2.5
1.0	0.0	−2.3
1.1	+0.3	−2.2
1.2	+0.7	−2.1
1.3	+1.1	−1.9
1.4	+1.7	−1.8

13 Fig. 9 shows a diagram of a kidney tubule.
    (a) Identify the structures A to H in the diagram.
    (b) Which of the following statements about kidneys are correct?
    (i) There are at least 1 000 000 tubules in a human kidney.
    (ii) Ultrafiltration occurs at the glomerulus under the influence of the enzyme rennin.
    (iii) The descending limb is impermeable to outward diffusion of water and ions.
    (iv) Some Na^+ ions passively enter the descending limb.
    (v) Some of the shortest ascending and descending limbs are found in animals that live in deserts, such as kangaroo rats.

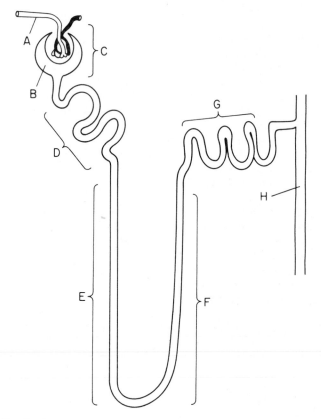

Fig. 9 Diagram of a kidney tubule

(vi) The ascending limb, under the influence of aldosterone, is permeable to the outward active transport of Na^+ and Cl^- ions, but is impermeable to the outward movement of water.

(vii) At the same horizontal level in each kidney tubule, Na^+ concentrations in the descending limb exceed those in the ascending limb, maintained by a counter-current multiplier.

(viii) The volume of filtrate produced by each human kidney is approximately 120 cm^3 per minute.

(ix) The human kidneys re-absorb approximately 79/80 volumes of water.

14 The marine worm *Nereis diversicolor* was placed into 50% sea water for a period of 16 hours. At the same time a tubular length of semipermeable membrane (Visking tubing) was filled with undiluted sea water and placed with *Nereis* in 50% sea water.

Changes in mass were recorded at intervals of 2 hours over a period of 16 hours, and expressed as percentage increase in mass (see Table 6).

(a) Plot these results as a graph.

(b) Give a reasoned explanation of the changes in mass observed both in the Visking tubing and in *Nereis* over a period of 16 hours. It is known that *Nereis*, when immersed in 50% sea water, remains hypertonic.

Table 6

Time (h)	% increase in mass	
	Visking tubing	Nereis
0	0	0
2	40	28
4	41	29
6	46	29
8	47	28
10	47	27
12	48	26
14	48	25
16	48	25

15 (a) Define the term *molar solution*.

(b) Table 7 lists the molecular weights of five compounds.
Describe how you would prepare each of the following solutions.
(i) 500cm^3 M NaCl
(ii) 250cm^3 0.1 M urea
(iii) 150cm^3 0.1 M glucose
(iv) 25cm^3 0.5 M sucrose.

(c) If you were provided with each of the solutions prepared in (b), calculate the amount of distilled water that would have to be added to each solution to obtain a 0.05 M solution of each solute.

(d) Define the term *osmotic pressure*.
What is the osmotic pressure of a molar solution, and what are the consequences of this?

(e) Calculate, in atmospheres, the osmotic pressure of the following solutions of glucose.
(i) 0.8 M
(ii) 0.5 M
(iii) 0.25 M.

(f) A researcher finds that cells of a fruit, containing a mixture of glucose and fructose, are isotonic with a 0.27 M solution of NaCl.
(i) What is the approximate combined glucose and fructose content of 100 cm^3 fruit juice?
(ii) Explain the meaning of the term *isotonic*.

Table 7

Compound	Molecular weight
Sodium chloride	58.44
Urea	60.06
Glucose	180.16
Fructose	180.16
Sucrose	342.30

16 (a) Define *passive diffusion*.

(b) What factors determine the rate of diffusion across membranes?

(c) Where in plants and/or animals would you expect to find the following compounds entering or leaving cells by diffusion?
(i) water (ii) oxygen (iii) carbon dioxide
(iv) NH_4^+.

(d) What is exchange diffusion and where does it occur?

(e) Define *osmosis*.

(f) Where, in mammals, is osmosis of importance in maintaining an efficient flow of blood within the circulatory system?

(g) What is active transport?

(h) Where does active transport occur?

17 (a) List six elemental macronutrients required by green plants.

(b) List five elemental micronutrients required by green plants.

(c) (i) What are the symptoms of nitrogen deficiency in plants?
(ii) Why do plants require a supply of nitrogen?

(d) Name three processes by which mineral ions may enter plant roots.

(e) Apart from via root hairs, at what other sites may ions be absorbed by plants?

(f) Name the ions that plants may absorb from the soil in order to obtain each of the following elements.
(i) nitrogen
(ii) sulphur
(iii) phosphorus
(iv) iron.

(g) Roots of oat plants were placed into a nutrient solution. After several days, concentrations of five different ions were measured both in the root and in the external solution. The results obtained are given in Table 8. What conclusions do you draw from these results?

Table 8

Ion	Concentration Solution ($\mu mol/cm^3$)	Root ($\mu mol/g$)
K^+	25	73
Na^+	25	2
Ca^{2+}	700	1
NO^-	8	56
SO_4^{2-}	0.1	15

Model A

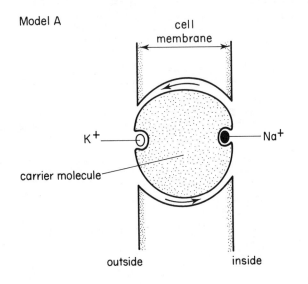

Model B

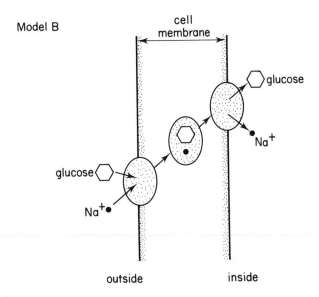

Fig. 10 Models of active transport across membranes

18 Fig. 10 illustrates two different models of active transport across cell membranes. What name has been given to each model, how is each believed to operate, and where might you find cells transporting materials by each of these mechanisms?

19 Indicate whether each of the following statements is true for xylem and/or phloem.
(i) Translocation may be acropetal, basipetal or bidirectional
(ii) It carries inorganic solutes
(iii) The conducting elements may become lignified
(iv) The conducting elements are associated with companion cells
(v) This tissue is investigated by the use of $^{14}CO_2$
(vi) Potassium (K^+) is mobile chiefly in this tissue
(vii) Translocation is currently explained in terms of the Cohesion–Tension theory.
(viii) Stained mostly red by fast green–safranin double staining.
(ix) Tyloses may block obsolete conducting elements in the tissue.
(x) Rates of translocation in this tissue may be influenced by light intensity.
(xi) Relative humidity may affect the rate of translocation in this tissue.
(xii) This tissue is not found in the gametophyte of a moss or fern.

20 The xylem forms a continuous system of vascular tissue throughout the plant.
(a) State the general functions of xylem.
(b) What differences can generally be observed
(i) between xylem vessels formed in spring and in autumn, and (ii) between xylem vessels located in the 'sapwood' and 'heartwood'?

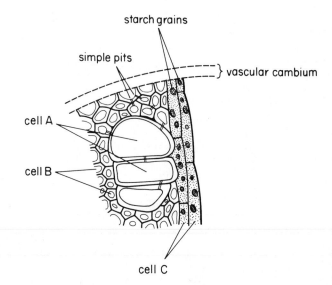

Fig. 11 Transverse section of cells of xylem in a woody stem

(c) Why is it incorrect to refer to water entering a xylem vessel by osmosis?

(d) What forces are believed to cause the flow of water through xylem? Are these forces active or passive?

(e) Fig. 11 shows the appearance of xylem in a transverse section of a woody stem. Identify the cells labelled A to C in the drawing, and make labelled drawings to show how these cells would appear in a longitudinal section of the xylem.

21 The model illustrated in Fig. 12 relates to a particular theory of translocation in the phloem.

(a) Write a brief account of the theory, relating it to the diagram.

(b) What are the shortcomings of the theory?

(c) What alternative proposals have been put forward?

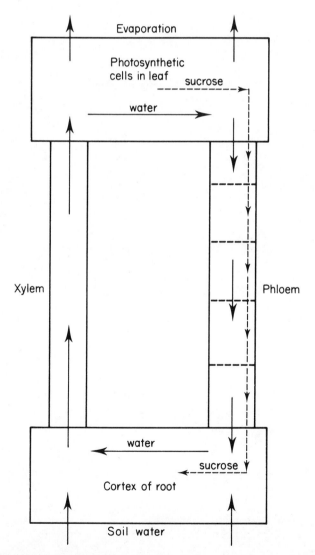

Fig. 12 A possible mechanism for the translocation of water and solutes through plants

22 Fig. 13 illustrates a simple Darwin Potometer.

(a) Make a complete drawing showing how a cut shoot would be positioned and held in place. Mark the position of the meniscus on your diagram.

(b) What precautions need to be taken when setting up a potometer of this type? What are its limitations?

(c) An investigator, working with a leafy shoot of oak (*Quercus robur*), investigated the effect of removing leaves on the rate of water uptake. The results obtained are given in Table 9. Plot these results as a graph.

Account for the small uptake of water after all of the leaves had been removed.

(d) A student, in a similar investigation, finds that the meniscus travels a distance of 6.5 cm in 5 minutes. If the capillary tubing has a diameter of 2 mm and the total leaf surface area is 263 cm², show how the student might attempt to calculate water loss in $cm^3\ m^{-2}\ h^{-1}$. What assumptions would have to be made?

(e) What method could have been used to estimate the surface area of the leaves?

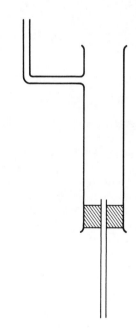

Fig. 13   Darwin's potometer

Table 9

No. leaves on shoot	Water uptake (cm³/hour)
6	3.6
5	3.1
4	2.4
3	1.9
2	1.4
1	0.9
0	0.3

23 Fig. 14 shows a diagrammatic representation of a stoma and its surrounding guard cells.
  (a) Briefly refer to the roles that have been assigned to factors A to F in the opening and closing of stomata.
  (b) Name the gases that would (i) be evolved from and (ii) enter an open stoma in bright sunlight.
  (c) List those factors, both internal and external, that would result in an increased rate of evolution of gases from stomata.

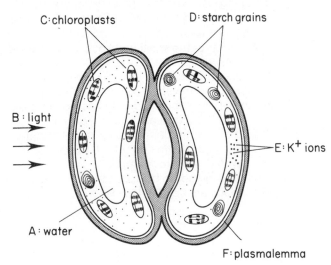

Fig. 14 Diagrammatic representation of a stoma and guard cells

24 Fig. 15 illustrates some components of mammalian blood.
  (a) Name the components A to E in the drawing.
  (b) What do the different components have in common with one another?
  (c) What difference would you observe if the blood sample had been taken from a frog or a bird?
  (d) Which of the cellular components are classified as granulocytes?
  (e) Which of the cellular components perform a phagocytic function?
  (f) With which cellular components are agglutinogens associated?
  (g) Which cellular components make the first response to invading antigens?
  (h) Of what significance is the biconcave shape of component A?
  (i) In which component would the chloride shift occur?
  (j) In which component would the Bohr effect take place?
  (k) Which cellular component is stored chiefly in the spleen?
  (l) Outline the chief function of component E.
  (m) Blood is a transport system, carrying a number of substances from one organ to another. From which organ, and to which organ, might blood transport each of the following compounds?

  (i) urea
  (ii) creatinine
  (iii) amino acids
  (iv) follicle stimulating hormone (FSH)
  (v) prothrombin.

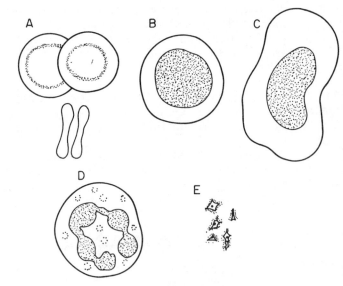

Fig. 15 Components of mammalian blood

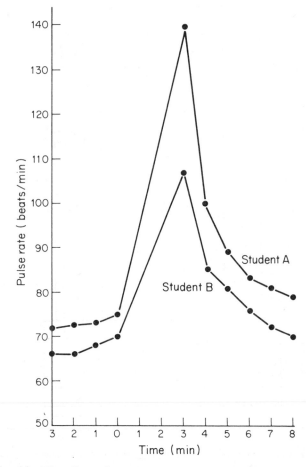

Fig. 16 The effect of a period of exercise on heart rate

25 Two students, A and B, went to a gymnasium and stepped on and off a bench for 3 minutes, in time with a metronome, set at 60 beats per minute. The pulse rate of each student was measured over a period of 3 minutes before the start of the exercise and for 5 minutes after its completion. Results are plotted in Fig. 16.

    (a) What is the pulse, and how would you measure pulse rates?

    (b) (i) What is the normal pulse rate in an adult human?

    (ii) Account for the slight rise in pulse rate recorded in the two students immediately before the exercise.

    (iii) In your view, which of the two students is the fitter? Give reasons for your answer.

    (iv) Criticise the design of the experiment and suggest ways in which more accurate results could have been obtained.

26 (a) Name, and briefly describe, the instrument used to measure blood pressure.

    (b) How is blood pressure measured?

    (c) What is the distinction between systolic and diastolic pressure, and how are these pressures measured?

    (d) Blood pressure is monitored by pressoreceptors and chemoreceptors positioned both in the aortic arch and carotid sinus. Sensory nerves lead from these receptors to the brain. Motor nerves run from the brain to the pacemaker of the heart and to circular muscles in the arteries, as illustrated in Fig. 17.

    (i) Explain the most probable sequence of events following a rise of blood pressure in the pressoreceptors.

    (ii) To which chemical stimuli are the chemo-receptors most likely to respond?

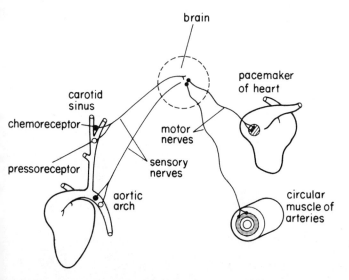

Fig. 17 The arrangement of receptors and effectors regulating blood pressure in a mammal

    (e) Consider Fig. 18, which shows three small blood vessels in transverse section. The vessels A, B and C have diameters in the ratio 1.0 : 0.75 : 0.5. If the peripheral resistance to blood flow in a vessel is proportional to the fourth power of its diameter, calculate the ratio of the resistance to blood flow ($1/d^4$) in the vessels A, B and C.

    (f) What relationship exists between cardiac output, arterial blood pressure and peripheral resistance?

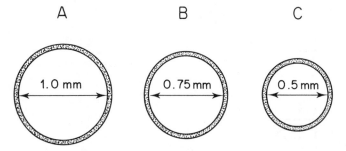

Fig. 18 Transverse sections of three small blood vessels

27 Indicate whether each of the following statements is true for arteries, veins and capillaries, or for one or more of these types of blood vessel.

    (i) They may carry deoxygenated blood.

    (ii) They carry blood from the heart to the tissues.

    (iii) They connect arterioles with venules.

    (iv) Their walls are permeable to glucose, amino acids and urea.

    (v) They offer the greatest resistance to blood flow.

    (vi) They contain smooth muscles in their walls.

    (vii) They return blood from the lungs to the heart.

    (viii) They are supplied with valves along their length.

    (ix) Constriction leads to an increase in stroke volume and cardiac output.

    (x) They are innervated mainly by the sympathetic system.

28 List the complete range of functions performed by water in the cells of plants and animals.
What changes would occur in herbaceous plants if the supply of water to their roots became intermittent?

29 How do terrestrial plants and animals conserve water?
In what ways are desert plants especially adapted for water conservation?

30 List the major ions, and trace elements, that are required by
    (a) vascular plants and
    (b) animals.
How do plants obtain their nutrients from the soil?
Describe how you would identify mineral-ion deficiency diseases in both plants and animals.

31 Three red blood corpuscles X, Y and Z leave the heart at the same time and embark on different journeys around the body. Fig. 19 shows some of the blood vessels through which they pass.

Complete the diagram by identifying blood vessels A to J.

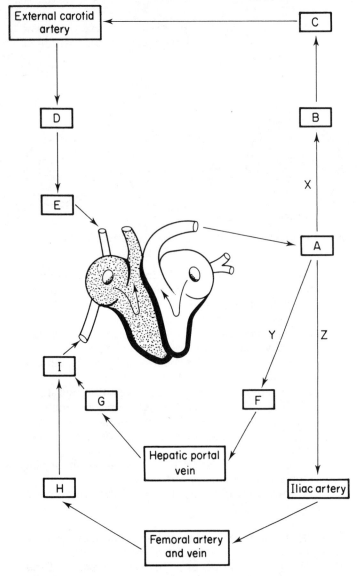

Fig. 19 Blood circulation through major blood vessels

32 Why do large, but not small animals, require transport systems?
What substances are carried in the transport systems of
(a) vascular plants and
(b) mammals?
Give an account of the mechanisms that are believed to effect transportation of solutes.

33 In a tall tree the leaves and roots may be separated by distances in excess of 30 metres. How does water pass from the roots to the leaves, and the products of photosynthesis pass from the leaves to the roots.

34 Make a large, labelled drawing of the heart of a mammal.
Describe the role of each functionally distinct region of the heart during a complete cardiac cycle.
Explain how variations in the rate of heart beat are brought about.

35 Compare and contrast differences in structure and functions that exist between
(a) arteries and veins,
(b) erythrocytes and leucocytes,
(c) blood and lymph,
(d) blood capillaries and lymph capillaries, and
(e) the left and right sides of the heart.

36 How would you determine the following experimentally?
(a) the osmotic pressure of epidermal cells of onion or rhubarb
(b) if yellowing of the leaves of a plant was caused by a mineral-ion deficiency
(c) if a shoot of cherry laurel transpired at a faster rate than a shoot of beech.

4 Modes of Nutrition and Excretion

Photosynthesis, feeding and excretion in animals, parasitism, saprophytism, mutualism

Multiple Choice Questions

1 The raw materials required for photosynthesis in vascular plants are
(a) carbon dioxide and oxygen
(b) carbon dioxide and water
(c) water and mineral salts
(d) carbon dioxide, water and mineral salts
(e) carbon dioxide, water, mineral salts and chlorophyll.

2 The optimum rate of photosynthesis in a vascular plant would be attained in
(a) blue light
(b) red light
(c) blue and red light
(d) green light
(e) white light.

3 Radioactive $^{14}CO_2$ is supplied to a photosynthesising mesophyll cell. Assuming that a sufficiently sensitive test became available, in which region of the mesophyll cell would you expect to detect the first traces of radioactivity?
(a) cell wall
(b) cytoplasm
(c) nucleus
(d) grana of chloroplasts
(e) stroma of chloroplasts.

4 An enzyme that plays an important role in CO_2 assimilation, and is associated with the wall of mesophyll cells, is
(a) phosphorylase
(b) hexokinase
(c) carbonic anhydrase
(d) glucose oxidase
(e) amylase.

5 The essential products of the light stage of photosynthesis, on which the dark stage is dependent, are
(a) CO_2 and H_2O
(b) ATP and $NADPH_2$
(c) ATP, $NADPH_2$ and CO_2
(d) ATP, $NADPH_2$ and O_2
(e) ATP, $NADPH_2$, CO_2 and O_2.

6 Chlorosis may occur in the leaves of a plant
(a) grown in darkness
(b) infected by a viral agent
(c) belonging to a particular genetic strain
(d) deficient in one or more chemical elements
(e) in any of the conditions listed in (a)–(d).

7 An excretory product retained within the cells of some vascular plants is
(a) carbon dioxide
(b) oxygen
(c) calcium oxalate
(d) calcium phosphate
(e) water.

8 The principal sources of lipase in the human alimentary canal is the
(a) stomach
(b) pancreas
(c) duodenum
(d) ileum
(e) liver.

9 Most of the amylase in the human alimentary canal is secreted by the
(a) salivary glands
(b) stomach
(c) pancreas
(d) duodenum
(e) ileum.

10 The following salivary glands are present in humans
(a) sublingual
(b) submaxillary·
(c) sublingual and submaxillary
(d) sublingual, submaxillary and parotid
(e) sublingual, submaxillary, parotid and infra-orbital.

11 Human skin is
(a) a major excretory organ
(b) a subsidiary excretory organ
(c) a rudimentary excretory organ
(d) an organ that plays no part in excretion
(e) an organ that plays no part in excretion or osmoregulation.

12 A human kidney excretes
(a) urea
(b) uric acid
(c) urea and uric acid
(d) urea, uric acid and creatinine
(e) urea, uric acid, creatinine and ammonia.

13 Uric acid is the major waste product resulting from the metabolism of proteins in
  (a) insects
  (b) insects and reptiles
  (c) insects, reptiles and birds
  (d) insects, reptiles, birds and mammals
  (e) reptiles, birds and mammals.

14 Bony marine fish excrete large amounts of
  (a) trimethylamine oxide
  (b) guanine
  (c) purines
  (d) allantoin
  (e) urea.

15 Mistletoe is
  (a) a saprophyte
  (b) a partial parasite
  (c) an obligate parasite
  (d) a temporary parasite
  (e) an ectoparasite.

16 Protective agents produced by the cells of green plants in response to fungal infections are called
  (a) phytochromes
  (b) phytoalexins
  (c) phytotoxins
  (d) antibiotics
  (e) fungicides.

17 In the life cycle of the malarial parasite (*Plasmodium*), which of the following stages might be found in a human host?
  (a) sporozoite
  (b) trophozoite and merozoite
  (c) trophozoite, merozoite and gametocyte
  (d) zygote and gametocyte
  (e) zygote.

18 In a newly discovered virus you might find
  (a) DNA
  (b) RNA
  (c) DNA or RNA
  (d) DNA and RNA
  (e) neither DNA nor RNA.

19 Which of the following plants benefits form an association with root nodule-forming bacteria?
  (a) onion
  (b) runner bean
  (c) cabbage
  (d) lettuce
  (e) carrot.

Short Structured Questions

20 The path of carbon dioxide in photosynthesis was investigated by Calvin and his associates, using a suspension of the alga *Chlorella*. A suspension of algal cells was allowed to photosynthesise in a stream of ordinary CO_2 in light. When photosynthesis was proceeding satisfactorily a portion of $H^{14}CO_3^-$ was injected into the system, and after a few seconds of photosynthesis with ^{14}C present the suspension of algae was run into hot methanol to denature proteins and to stop the reaction. Soluble materials were extracted, concentrated and chromatographed. The appearance of a completed radioautograph is illustrated, in a simplified form, in Fig. 20.
  (a) What conclusions can be drawn from the sequential appearance of compounds on the radioautograph?

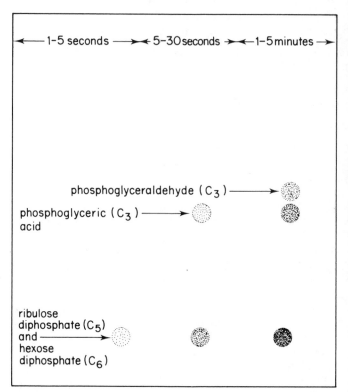

Fig. 20 Diagram of a radioautograph of products of the dark stage of photosynthesis

  (b) Fig. 21 shows the effect of O_2 concentrations on the rate of photosynthesis. What conclusions do you draw from this graph?
  (c) Fig. 22 shows the effects of CO_2 concentrations on the rate of photosynthesis. What conclusions may be drawn from the graph?
  (d) A potted plant, watered with $H_2^{18}O$, was placed in a CO_2-filled chamber and subjected to different intensities of illumination. Volumes of O_2 released from the plant were recorded (Table 10).

(i) Calculate intensities of illumination from the formula $1/d^2 \times 10\,000$, where d = distance in cm between the bulb and the plant. Plot your results as a graph.

(ii) Comment on the graph, indicating any way in which more satisfactory results might be obtained.

(iii) What would you conclude from the fact that the oxygen evolved was found to be radioactive?

(e) What factors, other than O_2 concentration, CO_2 concentration and light intensity are likely to influence the rate of photosynthesis in a vascular plant?

Table 10

Distance of plant from light bulb (cm)	Vol. O_2 evolved (cm^3/h)
80	0.1
60	0.3
40	0.9
20	1.6
10	3.2
5	53.5

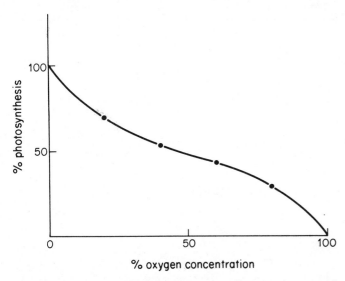

Fig. 21 The effect of oxygen concentration on the rate of photosynthesis

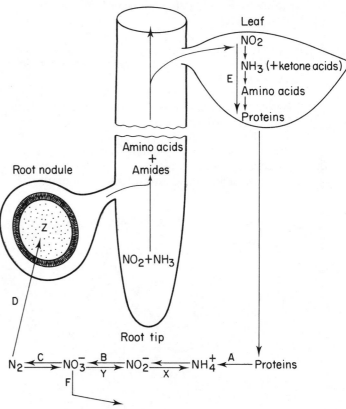

Fig. 23 The interconversion of nitrogen-containing compounds in green plants and in the soil

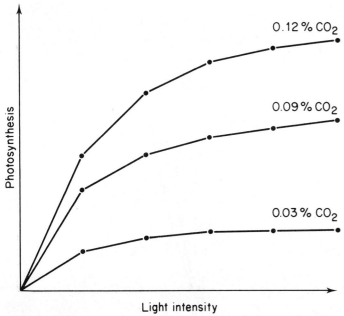

Fig. 22 The effect of carbon dioxide concentration on the rate of photosynthesis

21 Fig. 23 shows a diagrammatic representation of the root tip, nodule and leaf of a nodule-bearing legume, such as a bean or pea, together with compounds of nitrogen that occur both in the tissues of the plant and in its immediate surroundings. Processes, including chemical reactions, that occur in the plant and in the soil are lettered A–F.

(a) What processes are indicated in the diagram by letters A–F?

(b) Give the generic names of the bacteria responsible for processes X, Y and Z.

(c) Name the compounds that enter
  (i) the root nodule
  (ii) the root hairs at the root tip.

(d) What type of association is believed to exist between the root nodule bacteria and the legume?

(e) The root nodule bacteria are surrounded by cells containing the red pigment leghaemoglobin. What is believed to be the function of this pigment?

(f) In which tissue are amino acids which are formed in the root transported through the stem?

(g) Where in the plant would you expect to find (i) nitrate reductase and (ii) nitrite reductase?

(h) Name the most likely source of keto-acids in the leaves.

(i) In addition to free-living bacteria in the soil, what other common soil organisms are capable of nitrogen fixation?

22 (a) What do you understand by the *compensation point* of a plant? What happens when this point is exceeded?

(b) The results obtained in an investigation of the effects of temperature on the formation and utilisation of sugars in the leaves of a potato plant are given in Table 11.
From these figures, calculate net gain or loss of sugars at each temperature. Plot your results, together with those in the table, as a graph.

(c) In potato, what are the optimum temperatures for (i) photosynthesis and (ii) respiration?

(d) At what temperature does the potato reach its compensation point?

(e) By reference to your graph, explain the terms *true photosynthesis* and *apparent photosynthesis*.

Table 11

Temperature (°C)	Rate of sugar formation or utilisation (mg/h)	
	Photosynthesis (High light intensity)	Respiration
0	0	2
10	36	4
15	42	5
20	72	6
25	80	8
30	48	12
40	12	18
50	0	30
60	0	18

Organ W (sectional view)

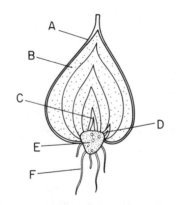

Organ X (surface view)

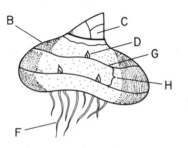

Organ Y (surface view)

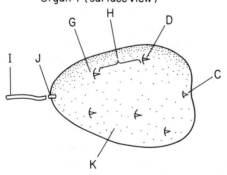

Organ Z (sectional view)

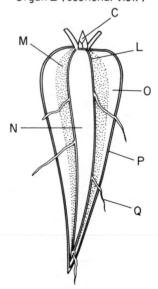

Fig. 24   Storage organs in herbaceous plants

23 A number of herbaceous plants store food reserves in modified leaves, stems or roots. The appearance of some of these organs, shown either in surface view or sectional view, is illustrated in Fig. 24.

(a) Name organs W–Z and identify the structures labelled A–Q.

(b) Complete Table 12, naming two examples of each type.

(c) What food reserves are stored in the body of a mammal and where are these areas of storage located?

Table 12

Storage organ	Examples	Nature of food reserves
W	(i) (ii)	
X	(i) (ii)	
Y	(i) (ii)	
Z	(i) (ii)	

24 The lower jaw of a dog is illustrated in Fig. 25.

(a) Identify the structural features A to H in the diagram.

(b) Make a large, labelled drawing of a vertical section through tooth D.

(c) In dogs, what functions are performed by the teeth labelled F, G, H and I?

(d) How do rabbits, or other herbivores, differ from the dog regarding the form and functions of their teeth?

(e) In addition to the form and arrangement of the teeth, what other differences generally exist between mammals that feed as (i) herbivores and (ii) carnivores?

(f) In dogs there are only two molar teeth in the upper jaw. Given this information, write the complete dental formula of a dog.

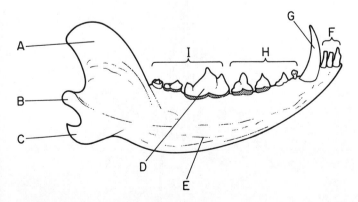

Fig. 25 Lower jaw of a dog

25 Fig. 26 shows a diagrammatic transverse section of the alimentary canal of a mammal.

(a) Identify the structural features A to I in the diagram.

(b) Which regions of the human alimentary canal are being described in the following statements?

(i) Regulates the flow of chyme from the stomach.

(ii) Brunner's glands are present.

(iii) Receives ducts from the parotid, sublingual and submandibular glands.

(iv) Secretion of enzymes occurs from Paneth cells.

(v) Oxyntic and peptic cells lie embedded within a cubical epithelium.

(vi) Microvilli line the inner border of columnar epithelium.

(vii) Water is absorbed from a region lined by simple columnar epithelium and goblet cells.

(viii) Ducts from the pancreas and liver pour their secretions into this region.

(ix) The most posterior sphincter muscle of the alimentary canal.

(x) Lined by stratified epithelium, perforated by ducts from mucus-secreting glands.

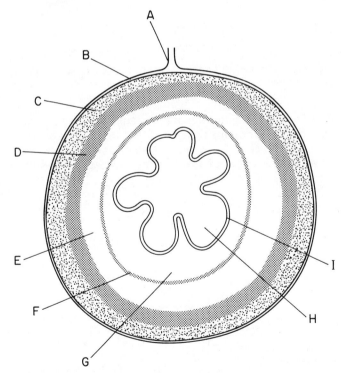

Fig. 26 Generalised transverse section of the alimentary canal of a mammal

26 (a) Distinguish between an *endopeptidase* and an *exopeptidase*, naming enzymes of each type that are active in the alimentary canal of a mammal.

(b) What feature, in mammals, characterises the secretion of endopeptidases?

(c) Fig. 27 is a summary of protein digestion in the alimentary canal of a mammal. Supply the missing items (i)–(v).

(d) Humans require nine essential amino acids in the diet: threonine, valine, methionine, leucine, isoleucine, lysine, histidine, phenylalanine and tryptophan. Explain why these amino acids are essential, while others are non-essential.

(e) What is the fate of excess amino acids present in the diet?

(f) What names are given to protein–calorie deficiencies in children.

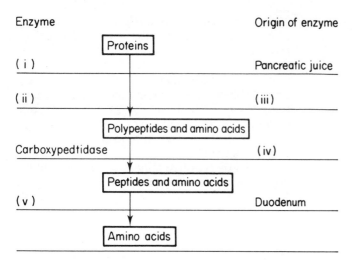

Fig. 27 Summary of protein digestion in the alimentary canal of a mammal

27 Fig. 28 illustrates successive regions of the alimentary canal. Hormones secreted by these regions are lettered A to G in the diagram. Name each hormone and give a brief account of its role in the process of digestion.

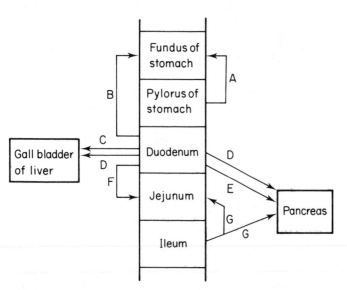

Fig. 28 Hormones produced by the stomach and small intestine

28   Indicate whether each of the following statements is applicable to
(a) an insect,
(b) a marine teleost fish or
(c) a mammal, or to one or more of these animals.
   (i) The excretory organs are kidneys.
   (ii) The excretory organs are malpighian tubules.
   (iii) These animals produce urine.
   (iv) The urine is generally hypertonic to the tissue fluids.
   (v) The urine contains some ammonia.
   (vi) Trimethylamine oxide (TMAO) may be excreted.
   (vii) Uric acid is the major excretory product.
   (viii) The kidneys have small or no glomeruli.
   (ix) The kidney tubules are long.
   (x) The excretory organs open into a part of the alimentary canal.

29 (a) Define the term *virus*.
   (b) Fig. 29 shows the structure of a bacteriophage. Name the structural features A to G.
   (c) What group of organisms is parasitised by bacteriophages?
   (d) Name a bacteriophage.
   (e) What are the two organic components of a virus?
   (f) What types of symmetry are exhibited by viruses?
   (g) What protective agent is synthesised by human cells in response to a viral infection?
   (h) List four ways in which viruses may pass from one plant host to another.
   (i) Outline the nature of the experiment, performed on a bacteriophage by Herschey and Chase, which showed that DNA regulates the synthesis of proteins.

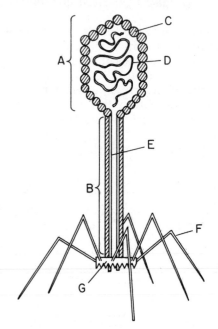

Fig. 29 A bacteriophage

30 Fig. 30 shows, in diagrammatic form, the structure of a bacterium.
- (a) Name the structures A to H in the diagram.
- (b) Bacteria are classified according to the shape of their cells. What forms are recognised?
- (c) Complete Table 13, which compares certain features found in bacterial cells with those found in the cells of higher plants.
- (d) List four different bacteria that are of economic importance, together with the product with which they are associated.
- (e) Distinguish between *autotrophic* and *heterotrophic* bacteria.

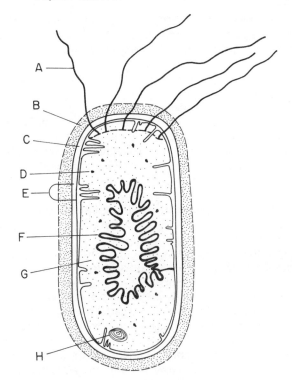

Fig. 30 Generalised structure of a bacterium

31 Which of the following diseases are caused by viruses and which by bacteria?
- (a) anthrax
- (b) poliomyelitis
- (c) tobacco mosaic
- (d) tuberculosis
- (e) rabies
- (f) cholera
- (g) peach leaf curl
- (h) chicken pox
- (i) typhoid
- (j) influenza.

32
- (a) Define the term *parasite*.
- (b) Distinguish, with examples, between
  - (i) endoparasites and ectoparasites
  - (ii) permanent and temporary parasites
  - (iii) facultative and obligate parasites.
- (c) What general methods are available to biologists for the control of parasites? Outline the principles involved in each method of control.
- (d) How do (i) pests and (ii) predators differ from parasites?

Table 13

Feature	Bacterial cells	Plant cells
(i) Cell walls		Composed of cellulose, hemicellulose and pectic acid.
(ii) DNA, Chromosomes and nucleus		DNA double-stranded, complexed with proteins into rod-shaped chromosomes. Numerous chromosomes are surrounded by a nuclear envelope.
(iii) Organisation of protoplasm		Ribosomes occupy a fixed position on the endoplasmic reticulum; organelles present.
(iv) Transfer of genes		Genes exchanged between cells in meiotic cross-over.

- (e) Name one or more chemical compounds currently used to effect control of the following pests or parasites.
  - (i) pathogenic bacteria in man and domestic animals
  - (ii) fungi
  - (iii) insects
  - (iv) molluscs
  - (v) weeds in lawns.
- (f) Animal parasites have developed certain biological characteristic which distinguish them from other animals. Some of these characters are listed below. In each case, name a parasite that shows the particular feature and describe, with reference to that parasite, the precise nature of the adaptation that is shown.
  - (i) reduction or loss of organs
  - (ii) increased reproductive capacity
  - (iii) modification of the life history
  - (iv) modification of existing structures and development of new structures
  - (v) physiological adaptations.

33
- (a) Wheat is an important cereal crop. The effect of applying nitrogen fertiliser on grain yield is shown in Table 14.
  State briefly the effects of increasing amounts of nitrogen fertiliser on each of the components numbered (i) to (iv).
- (b) Devise a formula that would enable a farmer to calculate grain yield in tonnes per hectare.
- (c) An experiment was conducted to investigate the effect of depth of fertiliser application on nitrogen uptake by wheat. Nitrogen fertiliser was applied at the rate of 150 kg/ha. Comment on the results given in Table 15.

(d) What compounds of nitrogen would you expect to find in an artificial fertiliser?

(e) A large amount of nitrogen fertiliser applied to the soil was not taken up by the plants. List some of the possible fates of this unabsorbed fraction of the fertiliser.

Table 14

	Nitrogen fertiliser applied (kg/ha)			
	0	50	100	150
(i) Grain yield (t/ha)	2.0	3.4	5.6	5.4
(ii) Ear population per metre of drill row	45	72	106	113
(iii) Ear size (grains/ear)	19	20	20	20
(iv) 1000 grains (mass(g))	36	38	38	36

Table 15

Depth of fertiliser application below surface (cm)	Uptake of nitrogen by wheat (kg/ha)
20	88
60	76
110	21
140	9

34 Fig. 31 shows the structure of a single hypha of a saprophytic mould fungus such as *Mucor*.

(a) Identify the structural features A to H.

(b) Enzymes such as carbohydrases, lipases and proteases are released from the hyphae of the fungus. What is the function of these enzymes and how does the fungus benefit from their activity?

(c) What role do mould fungi play in the economy of nature?

(d) How do certain fungi affect the growth of green vascular plants?

(e) What use does man make of fungi in industrial and other commercial processes?

Long Structured Questions

35 'All life is ultimately dependent upon energy from the sun.' Illustrate this statement by reference to the various modes of nutrition found amongst living organisms.

36 Give an account of the chief biochemical events that characterise

(a) the light stage and

(b) the dark stage of photosynthesis. To what extent are these two stages dependent upon one another? Is it possible to relate biochemical events in photosynthesis to the structural features of chloroplasts, as revealed by the electron microscope?

37 What are the constituents of a balanced diet? Make a large, labelled diagram of the alimentary canal of a named mammal.

Describe the digestion of starch and fat in a mammal. What is the fate of the products of this digestion?

38 Describe variations in the lining epithelium of different regions of the alimentary canal of a mammal. What is the significance of these variations?

39 Give an account of the variety and origin of excretory products in animals. How are excretory products eliminated from the bodies of animals?

40 Describe, with appropriate illustrations, stages in the life cycles of two parasites with definitive and intermediate hosts.

Explain how each of these parasites is transmitted from one host to another. How does a knowledge of the life cycles of the two parasites enable effective control to be achieved?

41 Write an essay on antibiotics.

42 Survey the commercial uses of bacteria and fungi.

43 What experimental methods would you employ to investigate the rate of photosynthesis in the following.

(a) an aquatic plant such as *Elodea*

(b) a potted plant such as *Primula*

(c) sun and shade leaves of ivy (*Hedera helix*), collected from a wood

(d) grass growing in a field or lawn.

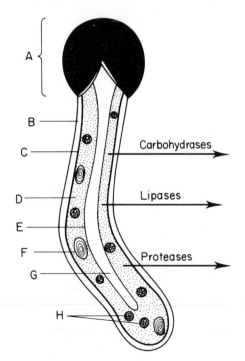

Fig. 31 Germinating hypha of a fungus such as *Mucor*

5 Respiration and Respiratory Systems

Tissue respiration, respiration in plants, respiration in animals

Multiple Choice Questions

1 The role played by ATP in biochemical reactions is that of
- (a) a reducing agent
- (b) an oxidising agent
- (c) an energy-donor substance
- (d) a co-enzyme
- (e) an energy-donor substance or a co-enzyme.

2 ATP synthesis occurs in
- (a) chloroplasts
- (b) mitochondria
- (c) chloroplasts and mitochondria
- (d) chloroplasts, mitochondria and nuclei
- (e) all cell organelles.

3 A compound found in mammalian skeletal muscle only that is, like ATP, a 'high energy' donor is
- (a) ADP
- (b) AMP
- (c) phosphocreatine
- (d) NAD
- (e) actin.

4 Details of the Krebs' cycle were worked out by the use of
- (a) X-ray crystallography
- (b) radioactive carbon compounds
- (c) ultracentrifugation
- (d) electron microscopy
- (e) light microscopy.

5 Which of these acids is an intermediate product in the Krebs' cycle?
- (a) hydrochloric
- (b) ascorbic
- (c) succinic
- (d) malonic
- (e) acetic.

6 During the Krebs' cycle a C_6 compound is formed by the chemical combination of a C_4 compound and a C_2 compound. The C_6 compound is citrate. Which are the C_2 and C_4 compounds involved?
- (a) malate and fumarate
- (b) acetyl CoA and oxaloacetate
- (c) isocitrate and succinate
- (d) acetyl CoA and fumarate
- (e) oxaloacetate and isocitrate.

7 In which region of a mitochondrion are enzymes of the Krebs' cycle believed to be located?
- (a) outer membrane
- (b) matrix
- (c) inner membrane
- (d) intermembrane space
- (e) outer membrane and inner membrane.

8 Iron-containing compounds that act as hydrogen acceptors in the respiratory chain are
- (a) flavoproteins
- (b) dehydrogenases
- (c) cytochromes
- (d) oxidases
- (e) anthocyanins.

9 In which of the following organisms would you *never* find mitochondria?
- (a) a bacterium
- (b) *Mucor*
- (c) *Spirogyra*
- (d) *Amoeba*
- (e) yeast.

10 The rate at which oxygen and carbon dioxide diffuse through a stoma is most directly related to the
- (a) radius of the stoma
- (b) circumference of the stoma
- (c) diameter of the stoma
- (d) total area of the stoma
- (e) size of the guard cells.

11 Gaseous diffusion through the stem epidermis of a herbaceous plant occurs through the
- (a) cuticle
- (b) stomata
- (c) stomata and lenticels
- (d) stomata and cuticle
- (e) lenticels.

12 The tissue that permits diffusion of gases through a lenticel is called
   (a) complementary
   (b) periderm
   (c) periblem
   (d) phellogen
   (e) plerome.

13 In which of the following regions of a root tip would you find cells with the highest rate of respiration?
   (a) root cap
   (b) meristem
   (c) root cap and meristem
   (d) region of elongation
   (e) region of differentiation.

14 If 1 g masses of the following compounds were oxidised, which one would require the largest amount of oxygen for complete oxidation?
   (a) protein
   (b) fat
   (c) glucose
   (d) sucrose
   (e) glycine.

15 In order to function efficiently, respiratory surfaces need to be
   (a) moist
   (b) thin
   (c) moist and thin
   (d) permeable to gases
   (e) moist, thin and permeable to gases.

16 During a 100 m sprint an athlete's respiration would be
   (a) entirely aerobic
   (b) aerobic, then anaerobic
   (c) anaerobic, then aerobic
   (d) entirely anaerobic
   (e) aerobic, anaerobic, then aerobic.

17 When a molecule of human haemoglobin is fully saturated with oxygen, how many molecules of oxygen does it carry?
   (a) 1
   (b) 2
   (c) 3
   (d) 4
   (e) 8.

18 The chloride shift refers to
   (a) the dissociation of $NaCl^-$ in the blood
   (b) exchange of HCO_3^- ions in the plasma for Cl^- ions in red corpuscles
   (c) exchange of HCO_3^- ions in the red corpuscles for Cl^- ions in the plasma
   (d) two-directional exchange of HCO_3^- and Cl^- ions between the red corpuscles and the plasma
   (e) a change in the percentage saturation of haemoglobin with oxygen.

19 Oxygen, passing from the atmosphere into the blood of a mammal, travels along the
   (a) bronchioles → bronchi → trachea → alveoli
   (b) alveoli → bronchioles → bronchi →trachea
   (c) trachea → bronchi → bronchioles → alveoli
   (d) trachea → bronchioles → bronchi → alveoli
   (e) bronchi → bronchioles → trachea → alveoli.

20 The carbohydrate most rapidly fermented by brewers' yeast (*S. cerevisiae*) is
   (a) glycogen
   (b) maltose
   (c) lactose
   (d) glucose
   (e) starch.

Short Structured Questions

21   (a) What does a biochemist understand by the terms *anaerobic* and *aerobic* respiration?
   (b) Where, in a cell capable of both (i) anaerobic and (ii) aerobic respiration, would these processes take place?

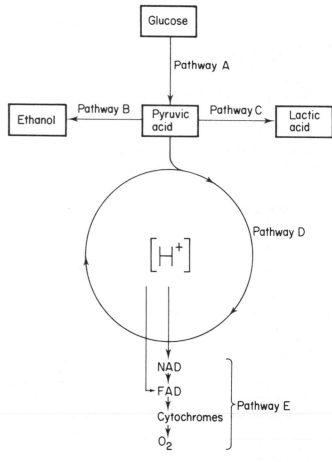

Fig. 32 Pathways by which a molecule of glucose may be respired

(c) A scheme, showing the aerobic respiration of a single molecule of glucose, is presented in Fig. 32. Name the metabolic pathways labelled A to E in the diagram.

(d) Name organisms in which metabolic pathways (i) B and (ii) C would occur.

(e) Each of the compounds in pathway E serves a similar function. What is that function?

(f) Name four acids formed as intermediate products in pathway D.

(g) From the complete oxidation of a molecule of glucose to CO_2 and H_2O, how many molecules of ATP are formed by (i) glycolysis and (ii) oxidative phosphorylation?

22 (a) What is the *respiratory quotient* of a respirable substrate?

(b) Calculate RQ values for each of the following equations:

(i) $C_6H_{12}O_6 + 6O_2 \rightarrow 6CO_2 + 6H_2O$

(ii) $C_6H_{12}O_6 \rightarrow 2C_2H_5OH + 2CO_2$

(iii) $C_{57}H_{104}O_6 + 83\ O_2 \rightarrow 57CO_2 + 52\ H_2O$
  glycerol
  trioleate

(iv) $C_6H_8O_7 + 4\frac{1}{2}O_2 \rightarrow 6CO_2 + 4H_2O$
  citric acid

(v) $2C_2H_2O_4 + O_2 \rightarrow 4CO_2 + 2H_2O$.
  malic acid

(c) Name two types of respirometer commonly used to measure RQ values in plant tissues.

(d) When using a respirometer, what compound would you use to absorb (i) O_2 and (ii) CO_2?

(e) Do you consider that RQ determinations give a reliable index of the nature of the substrate being respired? Give reasons for your answer.

(f) Table 16 shows, for humans, the amounts of oxygen required and the amount of energy released in the utilisation of different foods. Calculate the volume of CO_2 released from the complete oxidation of 1g (i) protein, (ii) fat and (iii) carbohydrate.

Table 16

Organic food	Oxygen required (litres)	Energy released (kJ/litre O_2 consumed)	Mean RQ value
1g protein	0.933	18.7	0.79
1g fat	1.989	19.7	0.71
1g carbohydrate	0.829	21.0	1.00

(g) What is the energy yield, in kJ, from the complete oxidation of 1g (i) protein, (ii) fat and (iii) carbohydrate?

(h) What is the smallest volume of O_2 that would have to be inspired in order to effect complete oxidation of a meal consisting of 25 g protein, 15 g fat and 150 g carbohydrate?

23 Fig. 33 shows the relationship between oxygen tension and the percentage oxygen saturation for human haemoglobin in an adult man at pH 7.5 and 7.2.

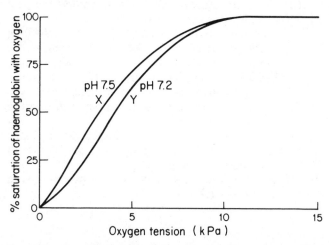

Fig. 33 The relationship between the percentage oxygen saturation for human haemoglobin in an adult man at pH 7.5 and pH 7.2

(a) What term is used to describe the shape of curves X and Y?

(b) Under what conditions might the pH of the blood change from 7.5 to 7.2?

(c) (i) What percentage of the total oxygen carried by the haemoglobin is released when the partial pressure of oxygen (the oxygen tension) 6.7 kPa at pH 7.5 and at pH 7.2?
(ii) What name is given to this effect, and what is its significance?
What other change in the body might produce a similar effect?

(d) (i) If atmospheric pressure were to fall to 86.7 kPa, what would be the oxygen tension?
(ii) What would be the percentage of saturation of haemoglobin at this oxygen tension?

(e) In which blood vessel would the percentage saturation of haemoglobin be (i) highest and (ii) lowest?

(f) To which side of curve X would you draw oxygen dissociation curves for (i) foetal haemoglobin, (ii) myoglobin and (iii) a haemoglobin from a mud-dwelling animal? Briefly explain your answers.

(g) Name two different blood pigments that occur in named invertebrate animals.

24 Fig. 34 illustrates structures concerned with the coordination of inspiratory and expiratory movements in a mammal.

  (a) Name the structures labelled A to E in the diagram.

  (b) Give a brief account of the role played by the labelled structures during inspiration and expiration.

  (c) What factors, acting on regions A and B, are likely to cause (i) an increased rate of ventilation, and (ii) a decreased rate of ventilation?

  (d) The consumption of oxygen during modest exercise is given in Table 17.
  Plot these results as a graph, *but extend the time scale to 15 minutes*. Add to the graph the predicted idealised fall in oxygen consumption between 7 and 15 minutes. What process is taking place during this time interval, and how is it related to the process that occurs during the period of exercise?

  (e) Which chemical factor, produced during exercise, is likely to have most effect on the rate of ventilation of the lungs?

  (f) What are the causes of muscle fatigue following vigorous exercise?

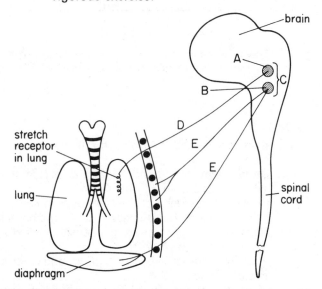

Fig. 34 Structures concerned with the coordination of inspiratory and expiratory movements in a mammal

Table 17

Time (min)	Activity	O_2 Consumption (cm^3/min)
0	rest	250
1	rest	250
2	exercise	630
3	exercise	815
4	exercise	890
5	exercise	1000
6	rest	not recorded
7	rest	not recorded

25 (a) For each of the following organisms, give the respiratory organs and state the method by which oxygen is conveyed to the tissues.
  (i) oak tree
  (ii) insect   (iv) fish
  (iii) earthworm   (v) frog.

  (b) Describe the mechanism of inspiration and expiration in a bony fish.

  (c) List the properties of respiratory membranes.

  (d) Why does an animal such as *Amoeba* not require a respiratory system?

  (e) What, in your view, is the chief difference between the way in which oxygen reaches the tissues in green plants and in vertebrates?

26 Fig. 35 shows a diagrammatic representation of the relationship between a blood capillary and air sacs of an alveolus.

  (a) If atmospheric pressure in the alveolus is 101.3 kPa, calculate the percentage composition of alveolar air by volume.

  (b) What is the source of most of the CO_2 and H_2O that enters the alveolus from the blood capillary?

  (c) In what form, or combination, is CO_2 carried from the tissues to the lungs?

  (d) What particular feature of red blood corpuscles do you associate with CO_2 transport?

  (e) (i) Comment on the shape and form of the alveolus in relation to gas-exchange.
  (ii) What would be the consequences of a breakdown in the structure of an alveolus as, for example, in emphysema?

  (f) Victims of drowning or suffocation, are generally given a mixture of 95% oxygen and 5% CO_2. What is the reason for this?

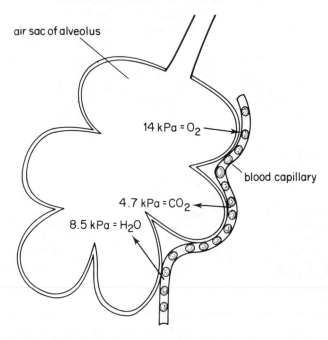

Fig. 35 Diagrammatic representation of the relationship between a blood capillary and the air sacs of an alveolus

27 (a) What instrument is used to measure lung volumes?
  (b) A spirometer trace was made of respiratory movements in a normal human subject over a period of 30 seconds. Results are shown in Table 18. Plot these results as a graph.
  (c) What was the subject asked to do at (i) 15 s (ii) 17.5 s and (iii) 25 s after the start of the recording?
  (d) Estimate from the graph
    (i) the vital capacity of the lungs
    (ii) the total volume of air inspired and expired in the first 15 seconds
    (iii) the residual volume of the lungs
    (iv) the tidal volume of the lungs.
  (e) If the same subject was asked to make a second recording while pedalling an exercise bicycle, what differences in the spirometer trace would be observed?

Table 18

Time (s)	Lung volume (l)
0	3.0
2.5	3.5
5.0	3.0
7.5	3.5
10.0	3.0
12.5	3.5
15.0	3.0
17.5	1.5
20.0	3.5
22.5	5.4
25.0	7.2
27.5	3.0
30.0	3.5

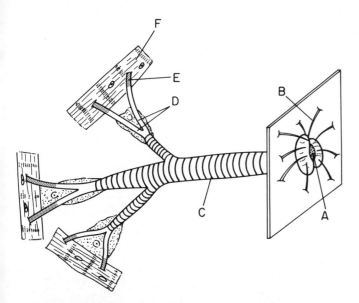

Fig. 36 Diagrammatic representation of a part of the tracheal system of an insect, including a surface view of an opening to a trachea in the cuticle

28 Fig. 36 shows a diagrammatic representation of a part of the tracheal system of an insect, including a surface view of an opening to a trachea in the cuticle.
  (a) Name the structures labelled A to F in the diagram.
  (b) What functions are performed by structures A and B?
  (c) Under what conditions might A be (i) open and (ii) closed?
  (d) Explain how oxygen reaches respiring cells in an insect.
  (e) Name the material that lines the tracheae. What happens to this material when an insect moults?
  (f) Explain why the possession of a tracheal system is generally considered to impose a limit on the size of insects.

29 Fermentation of glucose by yeast (*Saccharomyces cerevisiae*) may be represented by the statement:

$$C_6H_{12}O_6 \rightarrow 2C_2H_5OH + 2CO_2$$

Glucose   Ethanol   Carbon dioxide

(Atomic weights: C = 12, H = 1, O = 16)
If the Avogadro hypothesis is applied to this statement, it is possible to predict that the complete fermentation of one mole (180 g) glucose will yield 92 g ethanol and release 44.8 litres of CO_2 to the atmosphere.

Glucose	→	Ethanol	+	Carbon dioxide
180g		92g		44.8 litres

  (a) Calculate the theoretical yield of ethanol, by mass, when half-litre volumes of 0.5, 1.0, 1.5, 2.0, 2.5 and 3.0M solutions of glucose have undergone complete fermentation.
  (b) If the specific gravity of ethanol is 0.789 (sp. gr. water = 1.0), calculate the percentage of ethanol, by volume, that could be obtained from each glucose solution at the completion of fermentation.
  (c) Plot a graph showing theoretical yields of ethanol, both by mass and by volume.
  (d) Use your graph to find the mass of glucose, dissolved to make one litre of solution, that would be required to produce an alcoholic beverage containing
    (i) 5.7% ethanol by mass
    (ii) 11.3% ethanol by volume.
  (e) If you were making wine or beer at home,
    (i) what instrument would you use to measure the specific gravity of your fermentable substrate?
    (ii) Would you expect actual yields of ethanol to correspond exactly with predicted yields? Give reasons for your answer.
  (f) Some people make beer, or other alcoholic beverages, by allowing yeast and sugar to ferment in sealed jam jars or milk bottles. Why is this practice unsafe? What safety precautions should be taken?

Long Structured Questions

30 Explain concisely the essential features of
(a) anaerobic respiration and
(b) aerobic respiration.
Relate these two processes of respiration to structural features in regions of the cell where they are believed to take place.
What is the respiratory quotient (RQ), and what information does it provide about the nature of respiration that is occurring in a living organism?

31 Explain the importance of each of the following in the process of respiration
(a) respiratory membranes
(b) respiratory pigments
(c) cytochromes
(d) ATP   (e) oxygen   (f) temperature.

32 Distinguish between the processes of
(a) breathing and
(b) respiration.
Describe in detail the process of breathing in an insect, a fish and a mammal.
How are ventilation movements in a mammal co-ordinated?

33 What is yeast?
Describe the importance of yeast in (a) studies of respiration, and (b) brewing and baking.

34 What are fermentation processes?
Describe, by reference to at least five different examples, the economic importance of processes of fermentation.

35 How would you demonstrate the following experimentally?
(a) Germinating seeds can respire anaerobically
(b) Yeast can respire both anaerobically and aerobically
(c) People breathe out more CO_2 than they breathe in.

36 How would you measure the rate of respiration in the following organisms?
(a) a small invertebrate animal, such as a woodlouse
(b) a potted green plant, such as a geranium
(c) germinating seeds, such as those of pea or bean.

6 Reproduction, Development and Growth

Reproduction in plants, reproduction in animals, embryology and growth

Multiple Choice Questions

1 In flowers the term 'perianth segments' is applicable to
(a) sepals
(b) petals
(c) sepals and petals
(d) stamens
(e) stamens and carpels.

2 When the female gametes ripen first in a bisexual flower the condition is known as
(a) epigyny
(b) protandry
(c) protogyny
(d) perigyny
(e) hypogyny.

3 In which of the following would you find the haploid number of chromosomes?
(a) protonema of a moss
(b) prothallus of a fern
(c) pollen grain
(d) protonema, prothallus and pollen grain
(e) sporophyte of a moss.

4 If the generative nucleus of a pollen grain contains 12 chromosomes, how many chromosomes would you expect to find in a male gamete?
(a) 3
(b) 6
(c) 12
(d) 18
(e) 24.

5 The haploid number of chromosomes in a species of wheat is 7 and the diploid number is 14. How many chromosomes would you expect to find in cells of the endosperm?
(a) 7
(b) 14
(c) 21
(d) 28
(e) 35.

6 The acrosome of a spermatozoan is located in the
(a) head
(b) neck
(c) mid-piece
(d) nucleus
(e) tail.

7 A human egg is bounded by the
(a) jelly coat
(b) plasma membrane
(c) vitelline membrane
(d) plasma membrane and vitelline membrane
(e) plasma membrane, vitelline membrane and jelly coat.

8 When the offspring of an animal are born in a relatively advanced state of development, this is known as
(a) parthenogenesis
(b) parthenocarpy
(c) hermaphroditism
(d) syngamy
(e) viviparity.

9 Which of these hormones inhibits FSH production in a female mammal?
(a) oestrogen
(b) adrenalin
(c) thyroxin
(d) testosterone
(e) luteinising hormone.

10 Which of the following glands contributes a sugary secretion to human semen?
(a) testis
(b) prostate
(c) Cowper's
(d) seminal vesicles
(e) sebaceous.

11 Sexual reproduction always involves
(a) conjugation
(b) the fusion of gametes
(c) meiosis
(d) gametogenesis
(e) oogenesis.

12 The females of sharks, lizards, snakes and some insects, which retain yolk-filled eggs within the female reproductive tract for considerable periods, are described as
(a) oviparous
(b) viviparous
(c) ovoviviparous
(d) parthenogenetic
(e) microlecithal.

13 Which of the following has external fertilisation?
  (a) locust
  (b) frog
  (c) chicken
  (d) rabbit
  (e) man.

14 Budding is a method of asexual reproduction that occurs in
  (a) bacteria
  (b) *Spirogyra*
  (c) yeast
  (d) *Hydra*
  (e) yeast and *Hydra*.

15 Which of the following is hermaphrodite?
  (a) bee
  (b) earthworm
  (c) aphid
  (d) *Amoeba*
  (e) frog.

16 An example of an exalbuminous (non-endospermic) seed with epigeal germination is
  (a) maize
  (b) wheat
  (c) castor oil
  (d) broad bean
  (e) sunflower.

17 The region between the embryo and endosperm of a cereal grain is the
  (a) hilum
  (b) aleurone layer
  (c) coleoptile
  (d) coleorhiza
  (e) scutellum.

18 Which of the following is likely to exert most influence on the time of seed production in a flowering plant?
  (a) light intensity
  (b) light quality (colour)
  (c) periodicity of illumination
  (d) temperature
  (e) rainfall.

19 The archenteron, or primitive gut of a vertebrate embryo, develops during
  (a) fertilisation
  (b) cleavage
  (c) gastrulation
  (d) neurulation
  (e) organogenesis.

20 The neural plate of a vertebrate develops from the
  (a) neurectoderm
  (b) endoderm
  (c) mesoderm
  (d) archenteron
  (e) blastocoel.

21 Which records would provide the most informative data on the growth rate of individual plants in a field of wheat?
  (a) dry mass
  (b) wet mass
  (c) height
  (d) leaf area
  (e) root depth.

22 Which of the following plant hormones would promote the greatest rate of stem elongation in a dwarf pea plant?
  (a) gibberellin
  (b) abscisic acid
  (c) cytokinin
  (d) ethene (ethylene)
  (e) auxin.

Short Structured Questions

23 Complete the following passage by inserting appropriate words into the spaces lettered (a)–(n).

Individual plants bearing both unisexual male and female flowers are . . . (a) whereas plants that bear either unisexual male or female flowers are . . . (b) Flowers which shed their pollen before their stigmas are ripe are . . . (c) while those in which the stigmas ripen before the pollen are . . . (d) During fertilisation, the pollen tube enters the embryo sac via the . . . (e) and ruptures to release two . . . (f) One of these fuses with the . . . (g) forming a zygote from which the . . . (h) of the seed is formed. The second fuses with . . . (i) to form a triploid tissue, the . . . (j) Subsequent mitotic divisions of the zygote produces a . . . (k) from which a suspensor forms, and a . . . (l) cell, which develops into the cotyledons, . . . (m) and . . . (n) of the seed.

24 Fig. 37 shows the floral diagram of two flowers, each belonging to a different group of flowering plants.
  (a) If the ovary of each flower is superior, write the floral formula of each flower.
  (b) Make labelled drawings to show how each flower would appear if cut in half vertically.
  (c) What name is given to the arrangement of ovules in (i) flower A and (ii) flower B?
  (d) What terms are used to describe stamens that (i) open inwards, (ii) open outwards and (iii) protrude beyond the perianth segments and hang down?
  (e) To which groups of flowering plants do (i) flower A and (ii) flower B belong?

25 (a) Define the terms (i) *pollination* and (ii) *fertilisation*.
  (b) Name four different ways in which pollination may be effected.

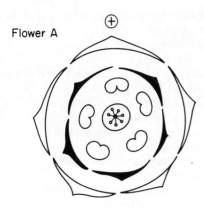

Flower A

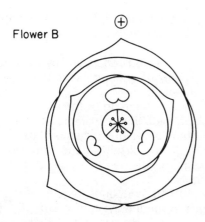

Flower B

Fig. 37 Floral diagrams of two flowers

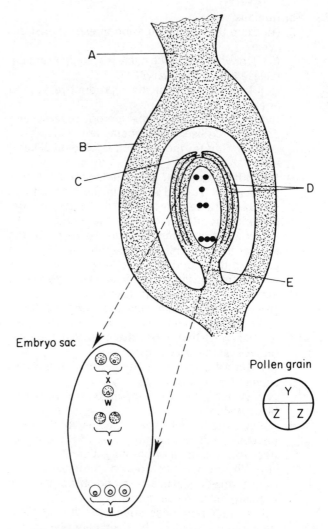

Embryo sac

Pollen grain

Fig. 38 Diagrammatic representation of a vertical section through an angiosperm ovary

(c) What are (i) anemophilous and (ii) entomophilous flowers?

(d) Name two differences that exist between the pollens of anemophilous and entomophilous flowers.

(e) What conditions must be met before a pollen grain will germinate?

(f) What process of cell division in a germinating pollen grain leads to the formation of two male gametes from the nucleus of a generative cell?

26 Fig. 38 illustrates a vertical section through an angiosperm ovary with a single ovule at the time of fertilisation. Beneath the ovary is an enlarged view of the embryo sac, together with the pollen grain that will effect fertilisation.

(a) Name the structural features labelled A to E and U to Z in the diagram.

(b) List, in their correct order, regions of the ovary through which the pollen tube must grow in order to reach the embryo sac.

(c) Following fertilisation, which components of the seed or fruit are formed from features labelled B, D, E, V and W.

27 Name two important examples of

(a) anemophilous flowers

(b) heterostylic flowers

(c) dioecious plants

(d) succulent fruits

(e) exalbuminous seeds

(f) albuminous seeds

(g) seedlings with epigeal germination

(h) seedlings with hypogeal germination

(i) vertebrates with external fertilisation

(j) hormones that regulate the menstrual cycle

(k) glands that contribute to the formation of semen

(l) insects reproducing by parthenogenesis

(m) insects with holometabolous metamorphosis

(n) embryonic membranes

(o) germ layers

(p) regions in an apical meristem

(q) plant growth hormones

(r) environmental factors affecting plant growth

(s) hormones that affect the growth of mammals.

28 For humans:
- (a) (i) State the normal duration of each menstrual cycle.
  (ii) On what day in the cycle is the egg normally released from the ovary?
  (iii) In which part of the reproductive system does fertilisation occur?
  (iv) In which part of the reproductive system does the fertilised egg become embedded?
  (v) Account for the production of identical and fraternal twins.
- (b) Fig. 39 illustrates an early human foetus. Name the features labelled A to G in the diagram.
- (c) (i) Name three compounds that pass, via the placenta, from the mother to the foetus.
  (ii) Name two compounds in the blood of the mother that are prevented by the placenta from entering the blood of the foetus.
  (iii) What features of the placenta enable a rapid exchange of materials to take place?
  (iv) The mammalian placenta performs functions for the foetus that, in an adult, are performed by distinct organ systems. What are the organ systems involved?
  (v) State the names, and functions, of hormones secreted by the placenta.
  (vi) Which group of mammals do not possess a placenta?
- (d) (i) What is the effective ingredient of the contraceptive pill, which suppresses FSH production from the anterior pituitary gland?
  (ii) Why are some doctors concerned about the long-term effect of this hormone on women who regularly take the contraceptive pill?
  (iii) Which hormones stimulate, during pregnancy, the growth of the mammary glands?
  (iv) What are the chief differences between teats and nipples?
  (v) Following from parturition, which hormone is responsible for the initiation of the milk flow?
  (vi) What is milk?
  (vii) Give a brief description of the function of rennin in the stomach of a young mammal.

29 In vertebrate embryology, what name is given to each of the following?
- (a) production of eggs
- (b) a fertilised egg
- (c) successive mitotic divisions of a fertilised egg
- (d) the future head end of an embryo
- (e) the central cavity of a blastula
- (f) cell movements within an embryo
- (g) the central cavity of a gastrula
- (h) the embryonic skeleton
- (i) the formation of a nerve cord
- (j) the outer germ layer
- (k) the fluid surrounding an embryo
- (l) regulators of differentiation
- (m) the membrane into which chick embryos void uric acid
- (n) a late human embryo
- (o) the influence of one developing tissue upon another
- (p) the point at which surface cells migrate into the interior of the embryo.

30 (a) (i) What do you understand by *metamorphosis?*
  (ii) Name a vertebrate that undergoes metamorphosis.
- (b) Explain the meaning of the terms
  (i) hemimetabolous metamorphosis
  (ii) holometabolous metamorphosis.
- (c) Give (i) one disadvantage and (ii) one advantage of the larval stage in the life of a butterfly.
- (d) Fig. 40 lists endocrine organs, hormones and stages in the life cycle of a butterfly. Copy the diagram and add arrows to indicate
  (i) the endocrine organs from which each hormone is secreted
  (ii) points at which hormones exert their influence.

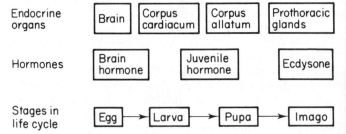

Fig. 40 Endocrine organs, hormones and stages in the life cycle of a butterfly

- (e) Fig. 41 illustrates graphs for the growth in size of an insect and the production of a hormone, which plays an important role in moulting.
  (i) What events take place at positions marked A, B and C in the diagram?
  (ii) What name could be given to events B and C, but not to A?
  (iii) State, with reasons, if the insect is hemimetabolous or holometabolous.

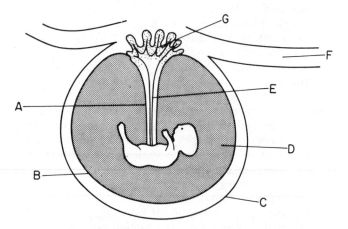

Fig. 39 Diagrammatic representation of an early human foetus

(iv) Name the hormone.
(v) Add to a copy of the graph, to the right-hand side of the growth curve, an additional curve to represent increase in mass.

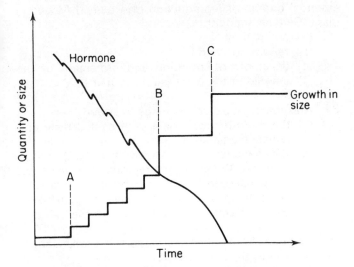

Fig. 41 Graph illustrating growth in size of an insect and the production of a hormone

31 Fig. 42 illustrates a vertical section through a maize grain.
   (a) Name the features labelled A to K in the diagram.

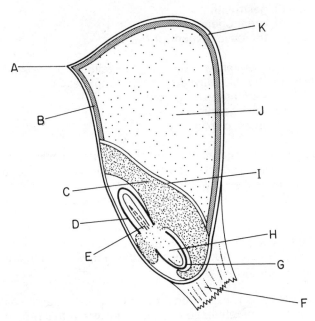

Fig. 42 Vertical section through a maize grain

   (b) Table 19 gives the changes in dry mass (g) during the first 10 days of germination of a maize grain.
   (i) Calculate changes in the dry mass of the grain.

(ii) Plot the results in the table, together with your calculated results, as a graph.
(c) Write a brief explanatory account of the process of germination in a cereal grain, using each of the following terms in your account. Underline each of the words the first time it is used.
Radicle, imbibition, starch, gibberellic acid, endosperm, plumule, glucose and fructose, maltose, enzyme induction, scutellum, α-amylase, embryo, aleurone layer.

Table 19

Time (days)	Dry mass (g)	
	Endosperm	Embryo
0	0.43	0.04
2	0.42	0.04
4	0.33	0.11
6	0.22	0.20
8	0.11	0.29
10	0.07	0.35

(d) Fig. 43 illustrates the seasonal growth of a maize plant.
   (i) How do you think the results were obtained?
   (ii) Describe the shape of the curve in the graph.
   (iii) What name is given to a graph of the seasonal growth of a plant such as maize?
   (iv) How do you account for the initial fall in dry mass of the seedling and the recovery that begins at the point lettered A in the graph?
   (v) How would you account for the final loss in mass from the point lettered B in the graph?

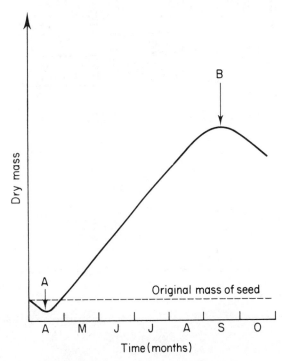

Fig. 43 Seasonal growth of a maize plant

32 Fig. 44 illustrates different types of life cycle found in living organisms.
  (a) Which of these life cycles occurs in each of the following organisms?
      (i) yeast
      (ii) a fern
      (iii) *Spirogyra*
      (iv) a flowering plant
      (v) humans.
  (b) Name one other organism, not listed above, that possesses a life cycle of (i) type A, (ii) type B and (iii) type C.
  (c) Write more detailed life cycles for any two organisms listed above, using the correct biological name or term to describe each stage in the life cycle.
  (d) What advantages do diploid organisms have over haploid ones?

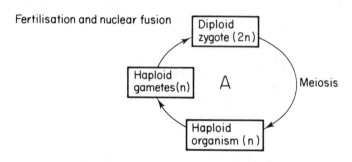

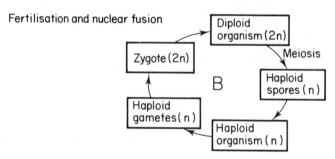

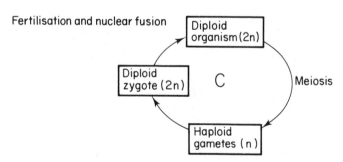

Fig. 44 Different types of life cycle found in living organisms

Long Structured Questions

33 Illustrate, by means of labelled diagrams only, the essential organs of reproduction of a named flowering plant. Give an account of
  (a) pollination,
  (b) fertilisation and
  (c) development, nutrition and protection of the embryo in the plant you have named.

34 Make large, labelled drawings to show
  (a) the reproductive organs of a named male and female mammal, and
  (b) the gametes of a named male and female mammal. Where, in the reproductive system of the female, do the gametes meet and fuse, and what changes occur in the fertilised egg before it becomes embedded in the wall of the uterus?

35 Compare, from a structural viewpoint, the similarities and differences that exist between
  (a) testis and ovary
  (b) entomophilous and anemophilous flowers
  (c) embryo sac and pollen grain
  (d) fern gametophyte and sporophyte
  (e) cleavage and gastrulation.

36 What is an embryonic organiser? Describe the role of organisers in
  (a) gastrulation,
  (b) organogeny and
  (c) the development of the eye.

37 With reference to auxin, gibberellin and cytokinin, describe the role played by plant hormones in
  (a) stem and root growth
  (b) seed germination
  (c) apical dominance.

38 Write illustrated notes on: the menstrual cycle, metamorphosis in insects, asexual reproduction, gametogenesis, hormones that influence the growth of mammals, parturition, embryonic membranes, germ layers.

39 Compare the process of fertilisation in a flowering plant and a mammal. What provisions exist for the protection and nourishment of the embryo in flowering plants and in mammals?

40 Write an essay on 'alternation of generations', making some reference to animals in which this process occurs.

41 How would you measure growth rates in each of the following?
  (a) a potted plant
  (b) a small mammal, such as a mouse
  (c) a population of bacteria
  (d) a standing crop of maize.

7 Support, Movement and Dispersal

Mechanical tissues in plants, bones and muscles in animals, mechanisms of dispersal

Multiple Choice Questions

1 The mechanical, or supporting, role in the stem of a herbaceous plant is performed by the
(a) parenchyma
(b) collenchyma
(c) fibres
(d) collenchyma and fibres
(e) parenchyma, collenchyma and fibres.

2 In a herbaceous plant, which of the following is supported chiefly by the water pressure of turgid parenchyma?
(a) leaf
(b) stem
(c) root
(d) stem and root
(e) fruit.

3 Cilia are used for locomotion in
(a) *Euglena*
(b) *Paramecium*
(c) *Vorticella*
(d) *Amoeba*
(e) *Plasmodium.*

4 The longest bone in the leg of a rat is the
(a) femur
(b) tibio-fibula
(c) humerus
(d) radius
(e) metatarsal.

5 An adult frog can move by
(a) running
(b) swimming
(c) swimming and jumping
(d) jumping
(e) jumping and running.

6 The skeletal element of a vertebrate embryo is the
(a) archenteron
(b) neural plate
(c) spinal cord
(d) vertebral column
(e) notochord.

7 Individual fibres of striated muscle are called
(a) myonemes
(b) myofibrils
(c) sarcomeres
(d) intercalated discs
(e) A-bands.

8 The length of myosin molecules in striated muscle determines the width of the
(a) I-band
(b) H-band
(c) A-band
(d) Z disc
(e) I-bands and A-bands.

9 The contractile proteins found in striated muscle are
(a) actin and myosin
(b) myosin and ATP
(c) actin and ATP
(d) actin and haemoglobin
(e) myosin and myoglobin.

10 Visceral (smooth) muscle does not occur in the
(a) alimentary canal
(b) arteries
(c) bladder
(d) eye
(e) heart.

11 The energy-donor substance(s) in striated muscle is/are
(a) ATP
(b) ATP and phosphocreatine
(c) glycogen
(d) glycogen and ATP
(e) glycogen and phosphocreatine.

12 Which of the following muscles are composed of individual cells?
(a) cardiac
(b) smooth
(c) cardiac and smooth
(d) striated
(e) striated and smooth.

13 Blood supplies muscles with
(a) oxygen
(b) ATP
(c) glycogen
(d) myoglobin
(e) energy.

14 Which of the following would accumulate as an end product of anaerobic respiration in a striated muscle cell?
(a) lactic acid
(b) malic acid
(c) citric acid
(d) succinic acid
(e) ethanol.

Short Structured Questions

15 Fig. 45 shows an idealised transverse section through the stem and root of a herbaceous plant.
(a) Name the tissues A to J in the diagram.
(b) What mechanical tissues are most likely to occur in the positions labelled W, X, Y and Z?
(c) Describe the arrangement of mechanical tissue in (i) the stem and (ii) the root. How does this arrangement enable the plant to withstand the effects of compression, extension and bending?
(d) List, in sequence, the principal events that occur during the secondary thickening of a stem.
(e) Of what importance is secondary thickening to a large plant, such as a tree?

Stem

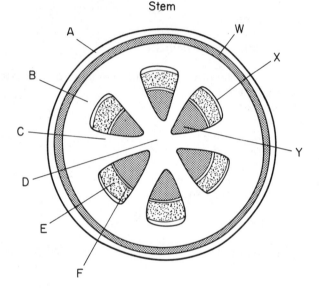

Root

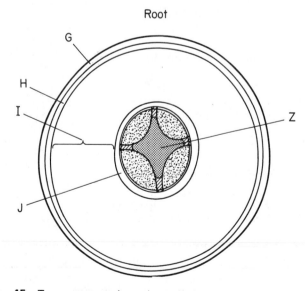

Fig. 45 Transverse sections through the stem and root of a herbaceous plant

16 Fig. 46 shows an outline diagram of a section through a long bone.

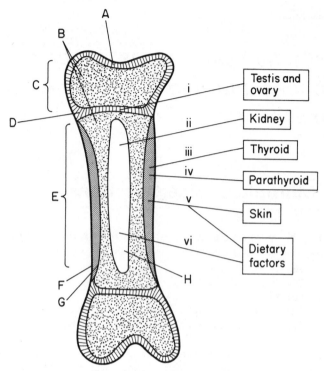

Fig. 46 Section through a long bone

(a) Name the structures A to H in the diagram
(b) The structure, form and functions of the long bones are determined by a number of factors, including chemical compounds, some produced by other organs and some that are ingested as an essential component of the diet. The origin of these compounds is indicated in Fig. 46, together with the region of bone on which they exert their principal effects. Identify the factors (i)–(vi) and write a short account of the effect of each factor on the bone.
(c) Fig. 47 illustrates the fine structure of an Haversian system.
(i) Name the structures labelled J to O in the diagram.
(ii) In which regions of the long bones would you expect to find Haversian systems?
(iii) What structures, in a living bone, would occupy spaces labelled L and M?

17 Fig. 48 shows a diagrammatic representation of the skeleton of a mammal.
(a) Name the features labelled A to N in the diagram.
(b) Clearly distinguish between (i) axial skeleton and (ii) appendicular skeleton. What distinction exists between (iii) cartilage bones and (iv) membrane bones?
(c) Make a large, labelled drawing to show the generalised structure of a vertebrate pentadactyl limb. Label this drawing with the names of bones

found in (i) the fore limb and (ii) the hind limb.
(d) How does the hind limb of a rabbit differ from the generalised pattern?

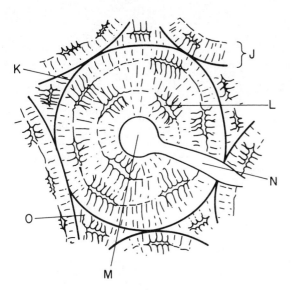

Fig. 47 Fine structure of an Haversian system

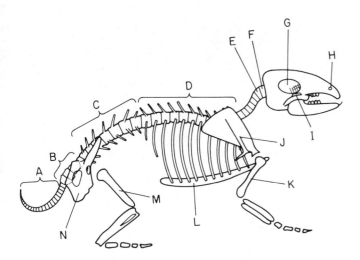

Fig. 48 Diagrammatic representation of the skeleton of a mammal

18 Fig. 49 shows a diagrammatic representation of two consecutive vertebrae in the spinal column of a mammal.
(a) Name the structural features labelled A to K in the diagram.
(b) What is the function of region J?
(c) What function is performed by muscles attached between consecutive regions labelled A–A' and B–B' on the diagram?
(d) List three functions of all vertebrae.

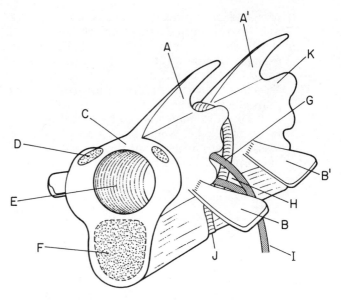

Fig. 49 Diagrammatic representation of two consecutive vertebrae in the vertebral column

19 Fig. 50 shows a vertical section through a movable joint.
(a) Name the structures labelled A to H in the diagram.
(b) What type of joints occur between bones of the mammalian skull? What is the function of these joints?
(c) Where, in the human body, would you find joints of the following types?
(i) simple sliding
(ii) compound sliding
(iii) rotating
(iv) ball and socket
(v) hinge.

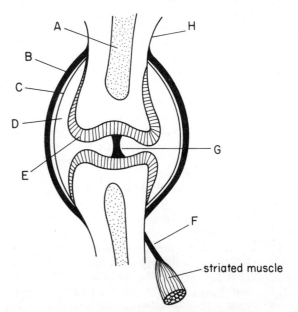

striated muscle

Fig. 50 Vertical section through a generalised movable joint

(d) Movable joints provide a fulcrum about which a pull from a muscle can effectively lift a load. Where, in the human body, would you find an example of a lever of each of the three types shown in Fig. 51.

(e) The mass of an animal is proportional to the cube of its length. If a small mammal 15 cm in length and 3.6 kg in mass were to be scaled up by a factor of 2, what would be its new length and mass? What effects might this increase in size have upon its bones and joints?

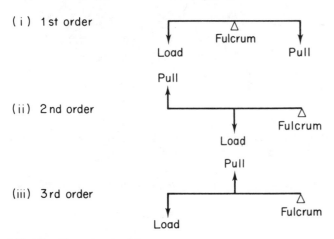

(i) 1st order

(ii) 2nd order

(iii) 3rd order

Fig. 51 Three types of lever

20 Which of the statements (a)–(j) relates to bone and which to cartilage?

(a) It joins the ventral end of the ribs to the sternum.

(b) Trabeculae may be present.

(c) The organic matter is ossein.

(d) It is always bounded by a fibrous membrane, the perichondrium.

(e) The mineral matter is a compound of calcium and phosphate.

(f) Secretory cells are positioned within lacunae.

(g) It is similar to dentine, but harder.

(h) It forms the skeleton in elasmobranch fish.

(i) Embryonic limb bones are pre-formed in this material.

(j) Cavities in this material are lined by a delicate membrane, the endosteum.

(k) Copy and complete Table 20, which relates to the distribution of cartilage within the human body:

21 Fig. 52 shows a vertical section through arthropod cuticle.

(a) Name the structures labelled A to I in the diagram.

(b) List three differences between the arthropod skeleton and the skeleton of a mammal.

(c) What properties of cuticle have, in your view, contributed most to the success of insects?

(d) What is the chief disadvantage of possessing an external cuticular skeleton?

Table 20

Type of cartilage	Location
Hyaline	Larynx (i) _____
	(ii) _____
Elastic	End of nose (iii) _____
	(iv) _____
(v) _____	Intervertebral discs

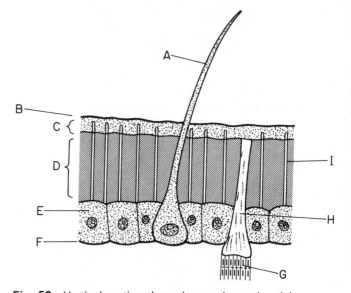

Fig. 52 Vertical section through an arthropod cuticle

(e) What name is given to moulting in insects? What hormone is associated with the process, and on what part of the cuticle does it exert its primary effects?

(f) Fig. 53 shows the foreleg of a bee. Name the principal segments of the limb, labelled J to N.

(g) What functions are performed by groups of hairs on the leg, labelled X and Y?

(h) In addition to the skeleton, what other important organs of insects are composed of chitin?

22 Fig. 54 shows a schematic representation illustrating the positioning of thick and thin filaments in striated muscles.

(a) Name the structures labelled A to G in the diagram.

(b) The diagram illustrates part of a striated muscle fibre in a relaxed position. What changes would occur when the muscle contracts?

(c) According to the sliding filament hypothesis, what determines the maximal shortening of the sarcomere that can be achieved during contraction?

(d) What is meant by the terms *isotonic* and *isometric* contraction?

(e) Striated muscle fibres are often described as *syncytial*. Explain the meaning of this term.

(f) In addition to striated muscle, what other types of muscle occur in the body of a mammal? How do these muscles differ from striated muscle?

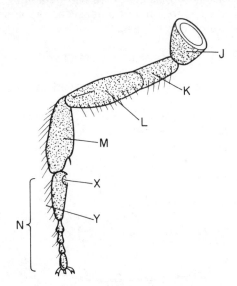

Fig. 53   Foreleg of a honey bee

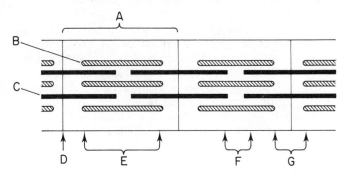

Fig. 54   Schematic representation of the positions of thick and thin filaments in a striated muscle fibre

23   Fig. 55 illustrates the principal muscles in the fore and hind limbs of a mammal.

(a) In the drawing of the fore limb, the brachialis muscle has been omitted. Insert this muscle on the drawing and describe its role in the functioning of the limb.

(b) Each muscle has a point of origin and a point of insertion. Explain the meaning of these terms. Indicate, with the letter O, points of origin of any three muscles on a copy of the diagram. Similarly, but in different muscles, use the letter I to indicate three points of insertion.

(c) What action will follow contraction of (i) the triceps muscle, (ii) the quadriceps muscle and (iii) the tibialis muscle?

(d) What do you understand by the statement that the biceps and triceps muscles are *antagonistic?*

(e) How is simultaneous contraction of the biceps and triceps muscles avoided?

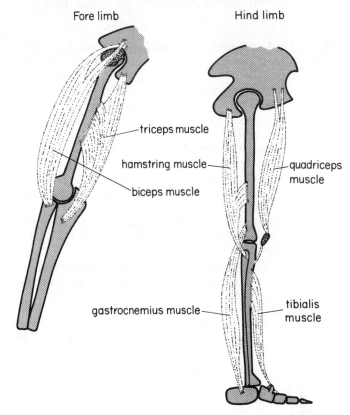

Fig. 55   Principal muscles in the fore and hind limbs of a mammal

24   Fig. 56 is a diagrammatic representation of a teleost fish, showing those structures that are of importance in swimming.

(a) Name the structures labelled A to G in the diagram. (Note that structure G is an internal organ.)

(b) What role is performed during swimming by structures labelled A and G?

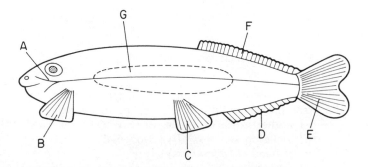

Fig. 56   Diagrammatic representation of a teleost fish

(c) What role is performed by each of the principal fins during swimming?

(d) How does (i) the shape of the body and (ii) the nature of the skin aid locomotion through the water?

(e) What additional problems are encountered by elasmobranch fish, such as dogfish and sharks, during swimming? How are these problems overcome?

25 Fig. 57 represents an insect in flight, with the lateral wall of the thorax cut away to show the muscles that control movements of the wing.

(a) On a copy of the diagram indicate the pattern traced by the wing tips during flight.

(b) Mark (i) the zone of lowest pressure (X) and (ii) the zone of highest pressure (Y) created by the wing movements during flight.

(c) Muscles of the thorax are not connected to the wings themselves. Attempt an explanation of the way in which these muscles effect up and down movements of the wings.

(d) During flight certain forces, indicated by the arrows labelled A, B, C and D, act on the insect. Identify these forces.

(e) What adaptations to flight are shown by insects?

(f) What additional adaptations to flight are shown by birds?

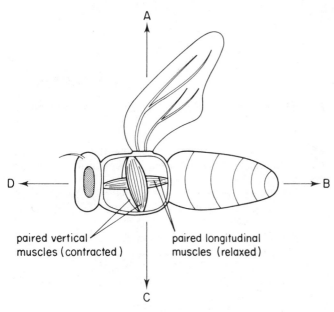

Fig. 57 Diagrammatic representation of an insect in flight

26 What vectors, or agents, are responsible for the dispersal, or distribution, of each of the following?

(a) antherozoids of a moss or fern

(b) sleeping sickness (*Trypanosoma*)

(c) malaria (*Plasmodium*)

(d) pollen grains from plants such as dead nettle, sweet pea and rose

(e) seeds and fruits of plants such as burdock and geum

(f) myxomatosis in rabbits

(g) rabies

(h) larvae of eels

(i) cholera and typhoid

(j) pollen grains of plants such as grasses, hazel and plantain

(k) viral diseases of plants

(l) genetic material from one bacterium to another

(m) seeds of balsam, gorse and geranium

(n) seeds and fruits of pine, sycamore and elm

(o) Dutch elm disease.

27 Fig. 58 illustrates an opened seed pod of gorse (*Ulex europaeus*).

(a) Name the structures labelled A to G in the drawing.

(b) The seed pod of gorse is covered by long, silky hairs. List two possible functions of these hairs.

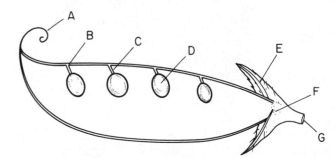

Fig. 58 An opened seed pod of gorse (*Ulex europaeus*)

(c) Seeds of gorse are dispersed by an explosive mechanism. The sound of an exploding pod can be heard several metres from a plant. Observations made over three successive days gave the results shown in Table 21. What do you conclude from these results?

(d) State how you would present the data in the table in a visual form. Give a reason for your answer.

Table 21

Time of day	Mean air temperature (°C) day 1	day 2	day 3	No. pods exploding day 1	day 2	day 3
0600–1200	16	23	20	0	8	6
1200–1800	19	26	25	0	21	18
1800–2400	17	19	19	0	0	0
2400–0600	15	18	18	0	0	0

(e) Seed pods of gorse were stripped from a stem and laid on a hot plate until they exploded. Results are given in Table 22. What do you conclude from these results?

(f) Name two other plants, growing in the wild, that disperse their seeds by an explosive mechanism.

Table 22

Distance of seed pod from stem apex (cm)	Mean time for pod to explode (s)
5	183
10	168
15	122
20	62
25	36

Long Structured Questions

28 Discuss, from a mechanical viewpoint, the problems that are created by the growth of large vascular plants, such as mature trees.
How are these problems solved, and to what extent do you consider the solution to be a compromise?

29 Discuss the view that the mammalian skeleton is an efficient piece of engineering.

30 Describe the location and structure of striated, smooth and cardiac muscle within the body of a named mammal.
Why does a mammal require more than one type of muscle?

31 By reference to a named quadruped mammal, describe the role played by its muscles and bones in (i) standing and (ii) running.

32 Review the modes and mechanisms of locomotion shown by fresh water protozoans.

33 Compare and contrast the flight of insects and birds.

34 What adaptations are shown by common inhabitants of ponds and lakes to (i) life on the surface of water, and (ii) life in an aquatic environment? Describe the various methods of locomotion shown by those freshwater invertebrates and vertebrates you have studied.

35 Give an account of the migrations undertaken by insects, fishes, birds and mammals. In what ways may these migrations by viewed as strategies for survival?

36 By reference to named viruses and bacteria, illustrate the role of dispersive agents in the spread of infectious diseases.

37 In what ways are vascular plants dependent upon dispersive agents for the successful completion of their life cycles? Illustrate your answer by reference to named examples.

38 Review the effects of wind, water and fire as agents affecting the success and dispersal of living organisms.

8 Stimuli, Responses and Behaviour

Tropisms and other responses in plants, photoperiodism, sense organs, nervous and hormonal responses in animals, behaviour

Multiple Choice Questions

1 Which is the correct sequence of events in the initiation of a tropic growth movement?
(a) stimulation→perception→transmission→response
(b) stimulation→transmission→perception→response
(c) perception→stimulation→transmission→response
(d) transmission→perception→stimulation→response
(e) response→perception→transmission→stimulation.

2 Cell organelles such as starch grains that probably provide the primary perception of gravity in plant cells are called
(a) klinostats
(b) kinocilia
(c) strobili
(d) statoliths
(e) statocysts.

3 The sudden collapse of leaflets of the sensitive plant *Mimosa pudica*, when they are touched or shaken, affords an example of
(a) thigmotropism
(b) hydrotropism
(c) epinasty
(d) seismonasty
(e) nyctinasty.

4 Long–short day plants, such as michaelmas daisy, require a period of long-day illumination followed by a period of short-day illumination before flowering will occur. At which season of the year would you expect these plants to flower?
(a) spring
(b) summer
(c) autumn
(d) winter
(e) spring and autumn.

5 Seeds of plants such as birch, hazel, apple and pine require a period of chilling before they will germinate. This chilling requirement is known as
(a) dormancy
(b) abscission
(c) scarification
(d) induction
(e) vernalisation.

6 In a relay (motor) neuron an impulse would pass, in sequence, along the
(a) axon → cell body → synaptic knobs → dendrites
(b) axon → dendrites → synaptic knobs → cell body
(c) dendrites → cell body → axon → synaptic knobs
(d) synaptic knobs → cell body → dendrites → axon
(e) axon → synaptic knobs → cell body → dendrites.

7 Which of the following is composed entirely of motor neurons?
(a) brain
(b) spinal cord
(c) somatic nervous system
(d) autonomic nervous system
(e) heart pacemaker system.

8 Which of the following statements is most generally applicable to the sympathetic nervous system?
(a) it causes contraction of smooth (plain) muscle
(b) it causes relaxation of smooth (plain) muscle
(c) it prepares the animal for activity
(d) its neurons secrete acetylcholine
(e) its neurons secrete noradrenalin.

9 Constriction of the pupil of the eye in bright light affords an example of
(a) sensory fatigue
(b) a spinal reflex
(c) a cranial reflex
(d) a displacement activity
(e) habituation.

10 Colour vision in vertebrates originates from stimulation of
(a) rods
(b) nerve cells
(c) cones
(d) chromatophores
(e) optic vesicles.

11 Which of these stimuli would evoke a response in a Pacinian corpuscle?
(a) light intensity
(b) gravity
(c) heat or cold
(d) sound
(e) pressure.

12 Receptors concerned with the sense of balance in mammals include the
(a) tympanum and cochlea
(b) tympanum and cristae
(c) cochlea and cristae
(d) cochlea and maculae
(e) cristae and maculae.

13 Which of the following sensations would be detected principally by taste buds at the back of the tongue?
(a) sweet
(b) bitter   (d) salt
(c) sour   (e) sweet and sour.

14 Which of the following *never* act(s) as an effector?
(a) cilia and flagella
(b) striated muscle
(c) exocrine organs
(d) endocrine organs
(e) neurons.

15 During pregnancy, the chief source of progesterone is the
(a) ovary
(b) placenta
(c) pituitary gland
(d) mammary glands
(e) uterus.

16 A rise in the level of blood glucose would normally occur following secretion of
(a) adrenalin
(b) thyroxin
(c) glucagon
(d) insulin and adrenalin
(e) glucagon and adrenalin.

17 A bird, reared in isolation from its parents, completed the task of nest-building at the first attempt. This is an example of
(a) instinct
(b) imprinting
(c) classical conditioning
(d) habituation
(e) insight.

18 The seasonal migration of birds affords an example of
(a) circadian rhythm
(b) circannual rhythm
(c) displacement activity
(d) taxis
(e) aestivation.

19 Dangerous animals, such as those which have poisonous bites, stings or obnoxious flesh are often
(a) white or colourless
(b) black or brown
(c) black and white
(d) striped or spotted
(e) marked with red, orange, yellow or blue.

20 Bird song is believed to
(a) communicate alarm or distress
(b) indicate flight intention
(c) maintain territory
(d) attract females and warn off other males
(e) perform all of the above functions.

21 When a number of stream-dwelling invertebrates are placed into a current of water, they will turn to face the source of the current. Which is the best term to describe this type of behaviour?
(a) tropotaxis
(b) klinotaxis
(c) chemotaxis
(d) rheotaxis
(e) geotaxis.

Short Structured Questions

22 A maize seedling with a straight plumule and radicle was subjected to unilateral illumination for a period of 12 hours. At intervals of 2 hours throughout the experiment, measurements were made to determine the angle of curvature of both the plumule and the radicle. The results are shown in Table 23.

Table 23

Length of exposure to unilateral illumination (h)	Angle of curvature	
	Plumule	Radicle
0	0°	0°
2	+7°	−3°
4	+14°	−4°
6	+18°	−8°
8	+23°	−11°

(a) (i) Plot these results as a graph.
    (ii) How do you interpret these results?
(b) What is a bioassay?
(c) How would you bioassay the amount of auxin in the coleoptile tip of a maize seedling?

(d) Fig. 59 shows the results of investigations into the effect of light on auxin production and distribution in excised coleoptile tips and coleoptile segments of maize. What interpretation would you place on these results?

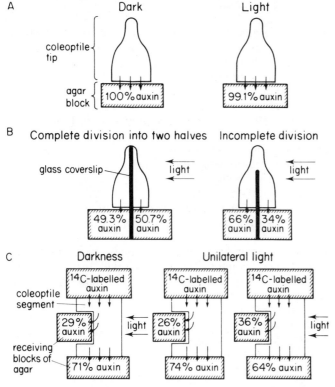

Fig. 59 The effects of light and darkness on auxin production and distribution in coleoptiles of maize. A shows continuous darkness and continuous light; B shows unilateral light, with the coleoptile tip either completely or partly subdivided by a glass coverslip; C shows unilateral light, with receiving blocks of agar placed both laterally and basally

(e) In a series of investigations into factors causing curvature of radicles, bean seeds were sown singly in plates of agar jelly. (The agar held the seeds firmly and supplied them with sufficient water for germination.) If the plates were supported with the jelly in a vertical position, all the emergent radicles grew downwards, over the surface of the agar, in response to gravity (see Fig. 60A).

What interpretations would you place on the results from each of the following treatments?
(i) The surface of the agar was covered by a polythene sheet. A small circular hole was cut in the polythene, to one side of the radicle tip. A total of 78 per cent of the radicles bent in the direction of the hole, the remaining 22 per cent showing no deviation from the vertical (see Fig. 60B).

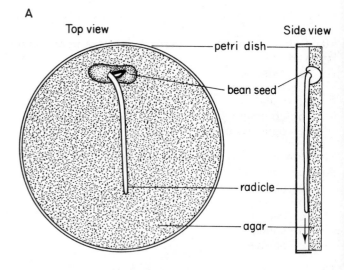

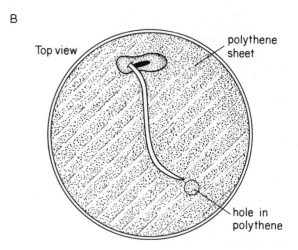

Fig. 60 Factors affecting bending of the radicle in a bean seedling. A shows the effect of gravity and B the effect of moisture

(ii) The test outlined above was repeated with seeds of other species, and the results are given in Table 24. How would you interpret these results?

Table 24

Plant species	% seedlings making positive response
Garden pea (*Pisum*)	0
Pine (*Pinus*)	0
Radish (*Raphanus*)	15
Onion (*Allium*)	23
Cucumber (*Cucurbita*)	73

(f) A researcher named Dodd grew seedlings of pea in plastic guttering, with holes bored through the guttering to allow the radicles to emerge. Beneath the guttering, and running parallel with it, a glass rod and wooden rod were attached, so that the emerging radicle was forced to grow between the parallel rods of glass and wood (see Fig. 61). Dodd observed that the radicles invariably grew in the direction of the wood. Suggest a number of alternative hypotheses that might account for this observation.

(g) How may the hydrotropic response differ from (i) the geotropic response and (ii) the phototropic response in radicles?

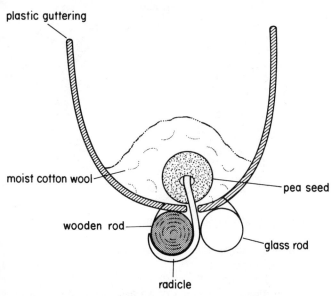

Fig. 61 Reconstruction of an experiment carried out by I. A. Dodd. All the radicles bent in the direction of the wood

23 Red and far-red light exert their effects on plants either by stimulating or by inhibiting certain processes.

(a) Copy and complete Table 25, stating whether the developmental process is stimulated or inhibited by red and far-red light.

Table 25

Developmental process	Effects	
	Red light	Far-red light
(i) Elongation of stems		
(ii) Expansion of leaves		
(iii) Growth of lateral roots		
(iv) Flowering of short-day plants		
(v) Flowering of long-day plants		
(vi) Germination of certain species of lettuce (e.g. Grand Rapids)		

(b) Name the photoreceptor pigment that absorbs red and far-red light, and state where the pigment would occur in a vascular plant.

(c) What physical properties of the photoreceptor pigment enable it to perform its biological functions?

(d) How is the pigment believed to influence flowering?

(e) Define, and give examples of
  (i) short-day plants
  (ii) long-day plants
  (iii) day-neutral plants.

(f) Plant A is a short-day plant and plant B is a long-day plant. Indicate, by placing a tick in the appropriate column of Table 26, the flowering response of each species under different periods of daylength.

Table 26

Daylength (h)		Flowering response	
Light	Dark	Species A	Species B
8	16		
7	17		
16	8		
17	7		

24 (a) What is homeostasis?

(b) What term is used of
  (i) animals such as mammals and birds that can maintain a constant internal body temperature.
  (ii) animals other than mammals and birds whose temperature equilibriates with the external temperature and shows some fluctuations.

(c) What changes occur in the skin of a mammal as an animal moves from
  (i) a cold region into a warm region
  (ii) a warm region into a cold region?

(d) What hormones are capable of stimulating heat output in a mammal?

(e) If a mammal is cold, what (i) involuntary and (ii) voluntary actions might be observed, both of which would increase heat production?

(f) Why are (i) babies and (ii) elderly people likely to lose heat more rapidly than adolescents or middle-aged members of the population?

(g) Blood glucose levels and the water volume of the blood are regulated by homeostatic mechanisms. Briefly describe the role of each of the following organs in these processes.
  Blood glucose levels:
  (i) islets of Langerhans
  (ii) adrenal medulla
  (iii) adrenal cortex
  Water and salt content of blood:
  (iv) anterior pituitary
  (v) adrenal cortex
  (vi) posterior pituitary.

25 (a) Fig. 62 illustrates a motor neuron. Name the features labelled A to I in the diagram.
(b) Which major sub-division of the nervous system is composed entirely of neurons of the same type as that illustrated in Fig. 62.
(c) Where, in the peripheral nervous system, would you find motor neurons?
(d) State two ways in which a sensory neuron would differ from a motor neuron.
(e) What do you understand by 'all or nothing' transmission of a nerve impulse in a single neuron?
(f) Briefly describe the events that occur at the points labelled H and I in the diagram that result in the transmission of an impulse to the next neuron in the series.
(g) (i) What is a neurotransmitter?
(ii) Name two neurotransmitters.

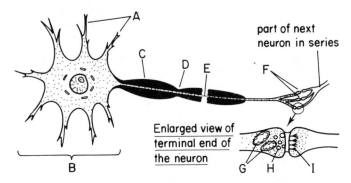

Fig. 62   A motor neuron

(h) Fig. 63 illustrates the arrangement of neurons across a part of the spinal cord. Name the features labelled J to P in the diagram.
(i) What name is given to the sequential arrangement of neurons J, K and L in the diagram? Is this on the left or right hand side of the body?
(j) Name the type of response that occurs when an impulse passes along this chain of neurons.
(k) Name two such responses, based on the spinal cord, that occur in mammals.
(l) Regions N and O are both composed of similar neurons, but each has a distinct colour. What are these colours and why do the two regions differ?

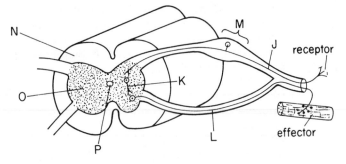

Fig. 63   Arrangement of neurons across a part of the spinal cord

26 Fig. 64 illustrates a vertical section through the eye of a mammal.
(a) Name the features labelled A to K in the diagram.
(b) Indicate, by adding appropriate lines to the drawing, the path of light rays from a nearby object to the retina in a person with
(i) faultless vision (label X)
(ii) myopia (Y)
(iii) hypermetropia (Z).

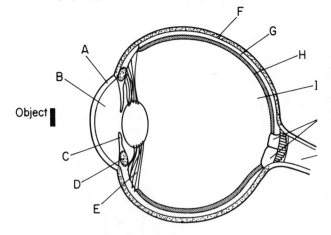

Fig. 64   Vertical section through the eye of a mammal

(c) Make clear, labelled drawings of
(i) a rod cell
(ii) a cone cell.
Indicate, by means of an arrow, the direction in which light reaches the visual pigment.
(d) Rods and cones perform different functions in the eye. To which of these two types of photoreceptor do the following statements refer?
(i) predominant in nocturnal animals
(ii) concentrated in and around the fovea of the eye
(iii) contain the pigment rhodopsin
(iv) contain the pigment iodopsin
(v) often arranged in groups, making contact with a single bipolar cell
(e) Fig. 65 illustrates four groups of receptor cells from the inner ear.

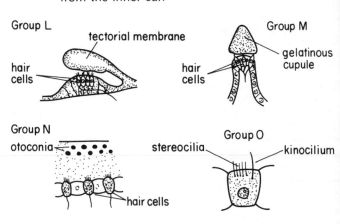

Fig. 65   Groups of receptor cells in the inner ear

For each of the diagrams labelled group L to group O
(i) name the part of the inner ear in which the receptor cells are located
(ii) name the stimulus to which the receptor cells respond
(iii) describe how the receptor cells are stimulated.

27 (a) Define the term *hormone*.
(b) Name two animal hormones that are (i) peptides or proteins and (ii) steroids.
(c) Name (i) two animal hormones that are antagonistic in their effects, and (ii) two plant hormones that are antagonistic in their effects.
(d) Complete Table 27, naming the endocrine organ in which each of the following animal hormones is produced, and the target organ on which it exerts its primary effect.
(e) What common feature do you associate with the secretion of vasopressin, ADH and oxytocin from the posterior lobe of the pituitary gland?
(f) Plant hormones have a number of commercial uses.
Copy and complete Table 28 by listing the use or purpose of each treatment.

Table 27

Hormone	Endocrine organ	Target organ
(i) Luteinising hormone (LH)		
(ii) Oxytocin		
(iii) Thyroid stimulating hormone (T.S.H.)		
(iv) Erythropoietin		
(v) Secretin		

Table 28

Plant hormone	Plant crop treated	Use or purpose of treatment
(i) IAA	Apple and pear orchards	
(ii) IAA	Woody cuttings	
(iii) 2,4-D	Lawns	
(iv) Gibberellin	Seed potatoes	
(v) Gibberellin	Black grapes	
(vi) Gibberellin	Barley (malting)	

28 Name *two* important examples of each of the following:
(a) spinal reflexes
(b) sign stimuli
(c) social insects
(d) sun-compass orientation

(e) photoreceptors
(f) mechanoreceptors
(g) nastic responses
(h) tropisms
(i) long-day plants
(j) short-day plants.

29 Courtship and mating in the three-spined stickleback fish follows a ritualised form, in which a stimulus from the male fish is followed by a response from the female. Next, the female stimulates the male, and this pattern continues until the male emits semen over the eggs. The various stimuli and responses are listed below, but not in their sequential order.
(a) vacates nest
(b) enters nest
(c) spawns; vacates nest
(d) approaches female; performs zig-zag dance three or more times
(e) follows male to nest
(f) stimulates spawning by prodding tail of female
(g) turns; leads to nest
(h) raises head; exposes swollen abdomen
(i) shows female entrance to nest; may dig floor of nest if female cannot be accommodated
(j) enters nest; emits semen.
Beginning with the approach of the male fish, re-list these stimuli/responses in their correct order, against either the male or female partner.

30 Name the behavioural response that is described in each of the following passages.
(a) Gull chicks peck at the tip of the parent's bill where there is a red spot. At this moment the parent regurgitates food and presents some of it to the chick.
(b) A nematode worm, (*Loa loa*), which infects people on Pacific islands, lives in the deep circulation at night and in the peripheral circulation in the daytime.
(c) Cows that were initially terrified by a passing train, soon learn to ignore it.
(d) Mice, placed into a maze, soon learn the most direct route to the cheese.
(e) Dogs, that normally salivate when they see food, were trained to salivate at the sound of a bell.
(f) Young ducks, reared in isolation from their parents, were observed to form an attachment to a research worker, and to follow him wherever he went.
(g) The chimpanzee piled up boxes to reach a banana hung from the ceiling.
(h) While waiting for an interview with the headmaster, the boy was observed to bite his nails and smooth his hair.
(i) During web-building an Indian spider (*Araneus nauticus*) lays a triangular frame of plain thread, runs radii across the web and then spins a sticky thread, crossing each radius three or four times. If parts of the plain thread are removed, leaving the

animal without any support between some radii, it does not repair the damage but struggles on with laying the viscid thread.

(j) Paths traced by two woodlice in a choice chamber were compared. The woodlouse in the dry chamber moved rapidly and turned more frequently than the one in the wet chamber.

(k) Certain birds such as the indigo bunting migrate chiefly at night, navigating by the position of the stars.

(l) If a bright light is shone into the eye, constriction of the pupil may generally be observed.

Long Structured Questions

31 Distinguish between (a) tropic and (b) nastic movements.
Outline the variety and importance of tropic and nastic movements made by green plants.

32 Write an essay on photoperiodism in plants and animals.

33 What mechanisms are believed to control each of the following processes?
(a) the geotropic response of roots
(b) the phototropic response of shoots
(c) the rate and stroke-volume of the heart
(d) levels of glucose in the blood.

34 What is homeostasis? Indicate the part played by each of the following in the homeostatic processes that occur in mammals: (a) hypothalamus (b) skin (c) diaphragm and intercostal muscles (d) pancreas.

35 Describe, under the following headings, the transmission of a nerve impulse from a receptor, such as a touch receptor in the skin, to an effector organ such as muscle:
(a) stimulation of the receptor
(b) transmission along neurons
(c) transmission between neurons
(d) response at a muscle.

36 Discuss the view that reflex arcs form the basic functional unit of the nervous system. Distinguish between
(a) cranial reflexes and spinal reflexes
(b) unconditioned and conditioned reflexes.

37 Make a large, labelled diagram to show the main regions of a mammalian brain. Briefly state the functions of the main parts of the brain of a mammal. Give a more detailed account of the functions of one region of the brain.

38 Describe the (a) origin, (b) structure and (c) functions of the retina. What changes occur in the eye during

(d) a sudden transition from darkness into bright light, and
(e) a change in focus from viewing a distant object to viewing one close at hand?

39 Make a large, clearly labelled drawing of the ear of a mammal.
Where, in the ear, are the principal sensory cells located? How do these sensory cells respond to stimuli such as sound, gravity, rotation and acceleration?

40 The pituitary gland has been described as the 'conductor of the endocrine orchestra'. Using specific examples of the hormones which it produces, and their effects on other endocrine organs, describe how the pituitary gland influences
(a) the basic metabolic rate (BMR) of the body
(b) the production of gametes
(c) the female menstrual cycle
(d) water retention and excretion
(e) birth and parturition in a female.

41 Attempt a classification of animal behaviour based on unlearned and learned responses.
Describe, with reference to named examples, the advantages and disadvantages associated with living in a society.

9 Ecology

Ecosystems, synecology, autecology, populations

1 Ecologists use different terms to describe regions or areas where organisms are found. Listed in order of decreasing size, these are
(a) territory → ecosystem → biosphere → habitat
(b) biosphere → ecosystem → habitat → territory
(c) territory → habitat → ecosystem → biosphere
(d) habitat → territory → ecosystem → biosphere
(e) biosphere → territory → habitat → ecosystem.

2 Two organisms in direct competition with one another are most likely to occupy the same
(a) territory
(b) niche
(c) trophic level
(d) microhabitat
(e) host.

3 Horizontal zonation of vegetation *always* occurs
(a) on a rocky sea shore
(b) in a pond
(c) in a field
(d) in a hedgerow
(e) in an oak wood.

4 The ecological growth efficiency of a herbivore may be expressed as
(a) $\dfrac{\text{production}}{\text{ingestion}}$
(b) $\dfrac{\text{egestion}}{\text{ingestion}}$
(c) $\dfrac{\text{production}}{\text{assimilation}}$
(d) $\dfrac{\text{ingestion}}{\text{assimilation}}$
(e) $\dfrac{\text{assimilation}}{\text{egestion}}$.

5 Which of the following would occupy the first trophic level in an oak wood?
(a) oak tree
(b) aphid
(c) wren
(d) squirrel
(e) fox.

6 In summer, which of the following habitats would show the most marked thermal stratification?
(a) puddle
(b) pond   (d) stream
(c) lake   (e) ocean.

7 On which of the following would you *not* recommend the use of a hormonal week killer, such as 2,4-D or MCPA?
(a) weeds in a lawn
(b) weeds in a field of wheat
(c) weeds surrounding a pond
(d) a gravel path
(e) weeds in an orchard of mature apple trees.

8 Which of the following would you expect to be absent, or rare, on very acid soils, deficient in calcium?
(a) protozoans
(b) snails
(c) insects
(d) calcifuge plants
(e) mosses.

9 The relationship between a single organism and its environment is its
(a) autecology
(b) synecology
(c) ecological valency
(d) ecological niche
(e) microhabitat.

10 Which of the following is not re-cycled within an ecosystem?
(a) water
(b) energy
(c) nitrogen
(d) carbon
(e) sulphur.

11 Roots of woodland trees such as beech, oak and pine, generally grow in symbiotic association with
(a) root nodule bacteria
(b) mould fungi
(c) ectotrophic mycorrhiza
(d) endotrophic mycorrhiza
(e) root eelworms.

12 Many plants growing in windswept, sandy upland regions show adaptations generally described as
(a) epiphytic
(b) parasitic
(c) xeromorphic
(d) mesophytic
(e) hydrophytic.

13 Which of the following would be used to estimate the percentage cover of daisies on a lawn?
  (a) line transect
  (b) belt transect
  (c) point transect
  (d) frame quadrat
  (e) point quadrat.

14 The expression $\dfrac{\text{total leaf area of plant}}{\text{area of ground covered by plant}}$ is known as the
  (a) biomass
  (b) unit leaf area
  (c) leaf area index
  (d) net primary production
  (e) gross primary production.

15 The hunting of lions, tigers and other animals to a point where their survival in the wild is threatened is known to biologists as
  (a) struggle for existence
  (b) over-exploitation
  (c) conservation
  (d) extinction
  (e) population management.

16 Which of the following pollutants is most likely to affect large numbers of plants growing by the sides of motorways?
  (a) sewage
  (b) lead
  (c) oil
  (d) mercurial salts
  (e) detergents.

17 The oxygen content of river water is lowered by the addition of
  (a) sewage
  (b) green plants
  (c) detergents
  (d) fertilisers
  (e) sewage, detergents and fertilisers.

18 Which of these pollutants, as they pass through food chains, are concentrated within the bodies of animals feeding at each trophic level?
  (a) insecticides such as DDT and BHC
  (b) caesium-137
  (c) insecticides such as DDT and BHC and caesium-137
  (d) herbicides such as diquat and paraquat
  (e) sewage.

19 Particle size 0.05–2.0 mm is a correct description of
  (a) gravel
  (b) clay
  (c) sand
  (d) loam
  (e) silt.

20 A podsol shows the following profile from top to bottom:
  (a) litter → bleached sand → grey clay
  (b) uniform brown horizon
  (c) organic matter → bleached sand → humus–iron pan → lighter stony layer
  (d) humus–iron pan
  (e) humus–iron pan → grey clay.

Short Structured Questions

21 (a) Read the following passage, which describes the flow of energy through an ecosystem, and then answer the questions that follow.

Solar radiation supplies the energy for primary producers. Only a small part of the total radiation (TR) received is available to the plants, the remainder (NU_1) is not used. The absorbed radiation (AR) is partly lost from the plants in the form of heat (H). The remainder is used in the synthesis of organic materials, and corresponds to the total photosynthesis or gross primary productivity (GP). The net primary productivity (NP) or apparent photosynthesis corresponds to the gross productivity less that quantity of organic material lost through respiration (R_1).

Part of net primary productivity (NP) serves as food for herbivores which assimilate an amount of energy I_1. Another portion of net primary productivity (NU_2) remains unused and passes, as dead plant material, to bacteria and other decomposers. The energy I_1 may be divided into two components, one of which A_1, is assimilated by herbivores while the other, NA_1 is discarded in the form of faeces and other wastes. Part of the assimilated energy (A_1) becomes secondary productivity (SP_1) and the remainder is expended in respiration (R_2). Finally, at the trophic level of the carnivores, assimilated energy (A_2), tertiary productivity (SP_2) and respiration (R_3) are similarly related to one another.

(Based on R. Dajoz, *Introduction to Ecology*, pp. 282–3, Hodder and Stoughton, 1977).

Write equations to represent
(i) gross primary productivity
(ii) net primary productivity
(iii) secondary productivity
(iv) energy flow across the herbivore trophic level
(v) the fraction of energy that is discarded by herbivores in the form of faeces
(vi) energy flow across the carnivore trophic level.

(b) Fig. 66 illustrates solar radiation falling on a leaf. What do the arrows (i), (ii), (iii) and (iv) represent, together with the percentage figures beside each arrow?

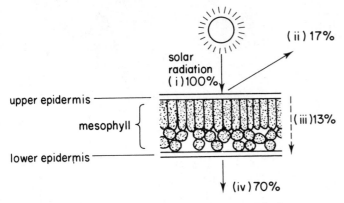

Fig. 66 The fate of solar radiation falling on a leaf

(c) Fig. 67 shows the energy budget of an ecosystem containing producers, hervivores, carnivores and decomposers. Figures in the diagram refer to kJ m^{-2} day^{-1}. Study the diagram and then answer the questions that follow.

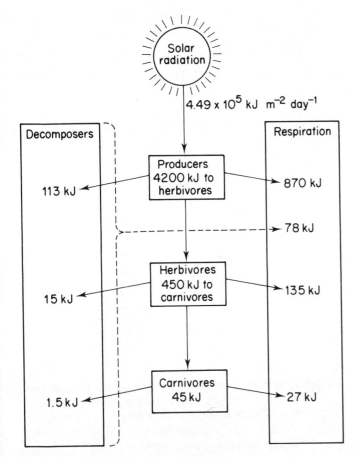

Fig. 67 The energy budget of an ecosystem

(i) Calculate the percentage of total incident radiation that is utilised by the producers.
(ii) What percentage of the gross energy fixed by producers is consumed by herbivores?
(iii) What percentage of the gross energy stored in herbivores is consumed by carnivores?
(iv) Calculate the net productivity of the producers.
(v) What percentage of the total annual assimilate is respired by the community?
(vi) Which group of organisms has the highest rate of respiration?
(vii) Which group of organisms contributes the largest percentage of their total biomass to the decomposers?

22 The following are factors that affect the distribution and population size of organisms. State which factors are climatic, which edaphic and which are biofic components of the environment.
  (a) light intensity
  (b) porosity
  (c) fire
  (d) annual rainfall
  (e) emigration
  (f) pH
  (g) fecundity
  (h) mortality
  (i) population density
  (j) natality
  (k) salinity
  (l) photoperiodism
  (m) relative humidity
  (n) competition.

23 Fig. 68 illustrates four different plant communities found on heathland throughout the British Isles.
  (a) Name (i) community A, (ii) community D and (iii) the gradual change in vegetation from A to D.
  (b) How would you determine the age of plants in communities B and C?
  (c) What environmental factors are most likely to cause
    (i) community B to revert to community A
    (ii) community C to revert to community B
    (iii) community D to be destroyed and revert to community A?
  (d) In which of the communities A to D would you find the following?
    (i) the largest number of insect pollinators
    (ii) the largest number of shrubs
    (iii) the largest number of species with wind-dispersed fruits and seeds
    (iv) most thalloid plants
    (v) soil with the highest water content
    (vi) the greatest net annual primary production
    (vii) the largest animal biomass
    (viii) plants most affected by grazing
    (ix) the simplest food webs.

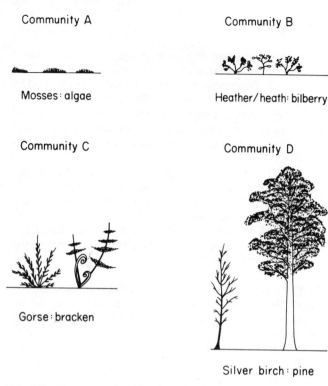

Community A

Mosses: algae

Community B

Heather/heath: bilberry

Community C

Community D

Gorse: bracken

Silver birch: pine

Fig. 68   Plant communities found on heathland

(e) Fig. 69 shows the usual duration of the leafy condition and the time of flowering for five of the species found in communities A to D. Identify, as far as possible, the five plants P to T. Explain how you arrived at your answers.

| Month |
Species	J	F	M	A	M	J	J	A	S	O	N	D
P												
Q												
R												
S												
T												

Key   ▒ flowers present   ▨ leaves present

Fig. 69   Leaf-bearing and flowering periods of five different plant species

24   Fig. 70 shows seasonal fluctuations in two adult insect populations which are interrelated.
(a) Describe the form of growth in each population.
(b) What is the most probable relationship between the two populations, and what role does each play in the relationship?
(c) What environmental factors are most likely to exert an effect on the numbers of individuals in population A?

(d) What do the levels K_1 and K_2 represent? What happens to a population if numbers (i) exceed and (ii) fall below these levels?
(e) How do population growth curves A and B differ from the growth curve for the human population?

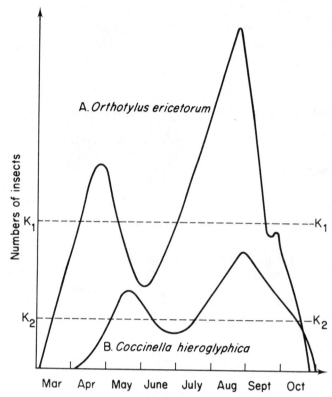

Fig. 70   Seasonal fluctuations in two interrelated insect populations

25   Table 29 records the growth of bacterial cells in laboratory culture.

Table 29

Time (h)	No. individuals in population
0	0
2	80
4	185
6	320
8	520
10	840
12	1025
14	1120
16	1180
18	1200

(a) Calculate, from the table above, the growth rate of the bacterium. Plot both the growth curve and the growth rate curve as a graph.
(b) What factors are responsible for limiting the size of the bacterial population?

(c) The figures in Table 30 are for populations of the flour beetle *Tribolium confusum*. Describe how population density is regulated in this species.

Table 30

No. adults per gram of flour	% of eggs eaten
1.25	7.7
2.5	17.0
5.0	20.0
10.0	29.7
20.0	70.2
40.0	98.4

(d) In the grain weevil *Sitophilus oryzae* experiments were carried out to determine the effect of population density on egg production. Results are recorded in Table 31.

Rewrite these figures so that they are comparable, and draw a graph to show the relationship between population density and egg production.

As population density increases, what factors may cause fewer eggs to be laid? (Figures based on R. Dajoz, *Introduction to Ecology*, Hodder and Stoughton, 1977.)

Table 31

No. female weevils	No. grains	Mean no. eggs laid per female
4	800	6.6
8	400	3.5
32	400	3.02
128	200	1.6
128	50	0.9

26 In a study of the distribution of seaweeds and periwinkles on a rocky sea shore, a line was stretched from low-tide to high-tide marks, and a counting frame used to determine the number of plants and animals at different levels across the shore (see Fig. 71).

(a) What name is given by ecologists to (i) the line (A) and (ii) the counting frame (B)?

(b) How would each of these pieces of apparatus be used in this survey?

(c) Before either piece of apparatus could be used, why would it be necessary to closely examine samples of the sea weeds and periwinkles and what special reference books would you need?

(d) Results from a survey of this type are presented in Table 32. Plot these results as a histogram to show any relationship that might exist between the distribution of seaweeds and periwinkles in different regions of the shore. What conclusions do you draw?

(e) You obtain results which suggests that the edible periwinkle (*Littorina littorea*) feeds exclusively on *Fucus serratus*. How would you investigate this hypothesis?

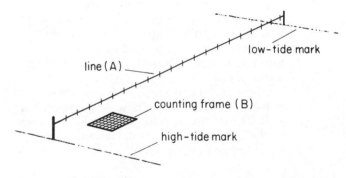

Fig. 71 Apparatus used to study the distribution of plants and animals on a sea shore

Table 32

Distance from low-water mark (m)	Number of individuals/m² shore							
	Sea weeds				Periwinkle			
	Fucus vesiculosus	*Ulva lactuca*	*Enteromorpha intestinalis*	*Fucus serratus*	*Littorina neritoides*	*Littorina rudis*	*Littorina obtusata*	*Littorina littorea*
0	—	—	—	—	—	—	—	1
2	—	—	—	21	—	—	—	5
4	—	2	—	26	—	—	1	8
6	—	5	—	13	—	—	3	3
8	2	10	—	2	—	3	6	2
10	6	12	—	—	—	2	3	1
12	12	8	—	—	—	8	3	1
14	14	3	10	—	—	3	1	—
16	1	—	10	—	3	1	1	—
18	—	—	—	—	1	—	—	—

27 A bean plant is cultivated for its annual yield of edible dried seeds. In a carefully controlled experiment, investigations were made into the effects of density of sowing on the number of pods per plant, the number of seeds per pod and the mean mass of the seeds. The results are shown in Table 33.

Table 33

No. plants/m²	No. pods/plant	No. seeds/pod	Mean mass of seed (g)
4	15.7	8.6	0.13
8	14.6	6.2	0.12
16	12.4	4.8	0.11
32	8.6	3.1	0.10
64	5.3	2.5	0.09
128	1.6	1.2	0.09

(a) Plot these results as a graph.
(b) (i) Calculate the total mass of dry seed harvested from each plot and the optimal density of sowing.
(ii) Why are densities below and above the optimal level likely to result in decreased yields?
(c) It was observed that at optimal density of sowing, the bean plants were particularly successful in overcoming competition from weed species.
(i) For what would the bean plants and weeds have been in competition?
(ii) What features possessed by the bean plants might have enabled them to be successful in this competition?
(d) Seeds of wheat and poppy were sown alone and as mixtures. Both sowings were made on bare earth. The results are given in Table 34. Comment on the effect of wheat on the growth of poppy, and attempt an explanation.

Table 34

	% population reaching flowering stage	
	Wheat	Poppy
Controls (each grown alone)	98.3	16.5
1 wheat: 1 poppy mixture	98.3	61.7

Table 35

	Mean dry mass/plant (g)		No. flowers/chick-weed plant
	Barley	Chickweed	
Controls (each grown alone)	5.15	3.16	117
1 barley: 1 chickweed	5.15	1.48	14

(e) Seeds of barley and chickweed were sown alone and as mixtures. After two months of growth plants were harvested and weighed, and counts made of the number of flowers per chickweed plant. The results are given in Table 35. Comment on these results and attempt an explanation.

28 A large shallow pond contains a population of diving beetles (*Dytiscus marginalis*). Five students, each making sweeps of the pond lasting for 3 minutes, removed a total of 58 beetles, which were marked and returned to the water. After allowing time for redistribution, further sweeps of the pond were made, and the results are given in Table 36. Captured beetles were not returned to the pond until after completion of the third sweep.

Table 36

Sweep	Total no. captured	No. marked individuals
1	143	16
2	136	16
3	150	19

(a) From these figures estimate the total number of diving beetles in the pond.
(b) Further sampling of the pond gave the results shown in Table 37. Plot these results in a way that will enable you to estimate the total population from the fewest samplings.

Table 37

Sample	No. captured	Total no. captured in previous samples
1	60	0
2	52	60
3	45	112
4	43	157
5	38	200
6	37	238
7	34	275

(c) An ecologist, using capture–recapture data, estimates that in a week a population declines from 500 to 400.
(i) What is the survival rate?
(ii) What is the death rate?
(iii) Criticise the term 'death rate' used in this context.

29 Fig. 72 maps the flow of a river through an area bordered by a number of sources of pollution, labelled A to F in the diagram.
(a) Name any pollutants that might enter the river from sources A to F, indicating the effect of each pollutant on the flora and fauna of the river.
(b) What single test could be carried out to provide an index of the degree to which the river water

was polluted? Give a brief description of the test.

(c) How would you determine if any raw sewage, containing faeces, had seeped into the river?

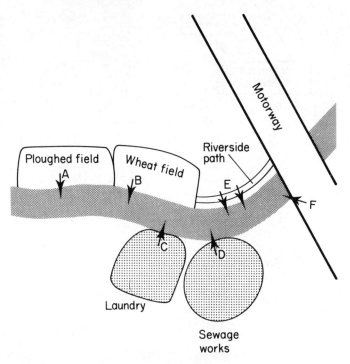

Fig. 72   Sources of pollution in a river

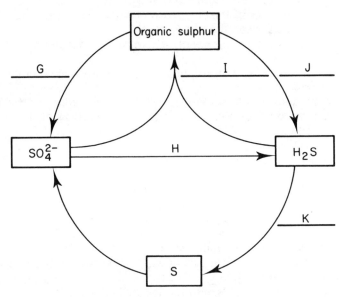

Fig. 73   Circulation of sulphur in a river

(d) Fig. 73 illustrates the circulation of sulphur as it might occur in rivers.
(i) Identify which of the following terms apply to the positions marked G to K in the diagram: mineralisation, oxidation, reduction, assimilation, putrefaction.

(ii) Which of the processes listed above are most likely to occur in a polluted river?
(iii) Name organisms that are most likely to assimilate SO_4^{2-}.
(iv) In what form would you expect to find sulphur in a well-aerated, fast-flowing river?
(v) Name two sulphur-containing compounds that occur in the body of a mammal.

30 (a) Percentages (dry mass) of the living and non-living components of a garden soil are listed in Table 38. Present this information in the form of three separate pie charts.

Table 38

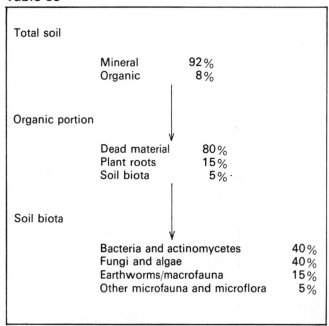

Total soil		
	Mineral	92%
	Organic	8%
Organic portion		
	Dead material	80%
	Plant roots	15%
	Soil biota	5%
Soil biota		
	Bacteria and actinomycetes	40%
	Fungi and algae	40%
	Earthworms/macrofauna	15%
	Other microfauna and microflora	5%

(b) Complete the entries in Table 39, which contrasts certain aspects of sand-based and clay-based soils.
(c) List the beneficial effects of earthworms in the soil.

Table 39

Property of soil	Sand	Clay
(i) Particle size		
(ii) Porosity		
(iii) Condition in winter		
(iv) Condition in summer		
(v) Mineral salt content		
(vi) Effect of wind		

(d) Fig. 74 shows the effect of burning wheat stubble on soil pH, measured at the surface.
(i) Account for the marked rise in pH following the fire.
(ii) What causes the pH slowly to return to its original level?
(iii) Many farmers now burn their wheat stubble following harvesting. What, in your view, are the advantages of this practice?
(iv) Does stubble-burning have any disadvantages, and, if so, what are they?

(e) Fig. 75 is an outline diagram of the nitrogen cycle. Copy and complete the diagram by adding one of the following terms, or names, to each of the positions marked (i) to (v) in the diagram: *Nitrosomonas, Nitrobacter*, ammonifiers, nitrogen fixers, denitrifiers.
(vi) List the nitrogen-containing organic residues produced by animals.

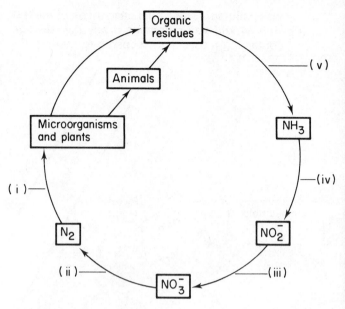

Fig. 75 An outline of the nitrogen cycle

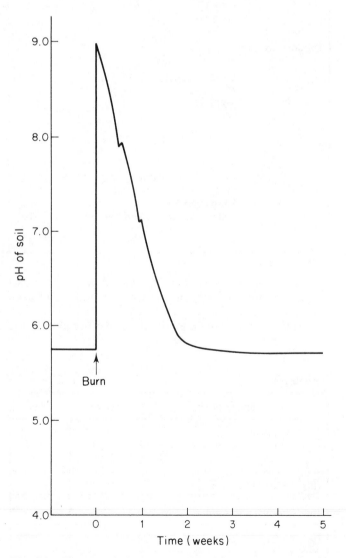

Fig. 74 Changes in the surface pH of the soil, following the burning of wheat stubble

Long Structured Questions

31 Choose two well-defined habitats. For each habitat you choose
(a) describe adaptations shown by the producers
(b) describe the diversity and interactions that occur between the consumers
(c) describe the variety and role of the decomposers.

32 Give an account of the autecology of a named plant or animal.
How does this autecology relate to the synecology of the habitat in which this plant or animal is generally found?

33 Distinguish between
(a) autecology and synecology
(b) microhabitat and macrohabitat
(c) population and community
(d) consumers and decomposers
(e) air pollutants and water pollutants.

34 What do you understand by the term *conservation?*
Account for the decline of, and argue the case in favour of conserving
(a) small fields, surrounded by hedges
(b) whales
(c) a *named* rare plant
(d) weeds of wheat fields *or* roadside verges
(e) large areas of mature woodland
(f) varieties of cultivated flowers and vegetables that have been replaced by 'improved' varieties.

35 How may field studies in ecology be complemented and extended by studies made in the laboratory?

36 Give a brief account of the carbon and mineral cycles in the soil. To what extent is the availability and uptake of mineral salts by plant roots influenced by
  (a) rainfall
  (b) activities of soil bacteria
  (c) soil pH
  (d) soil temperature?

37 *Either* write an essay on the use of statistical tests in ecological studies
or explain why field studies in ecology are considered to be of particular value in teaching scientific method.

38 Explain and discuss the ecosystem concept.

39 What (a) physical and (b) chemical properties of soil make it a suitable medium for the growth of plants? Describe the variety of soil types found in the British Isles. To what extent is the natural vegetation of (i) woodlands, (ii) chalk downlands and (iii) heathlands determined by the physical and chemical nature of the soils on which they occur?

40 What methods may be used to estimate the population density of plants and animals in particular habitats? What effect are (a) climatic, (b) biotic and (c) edaphic factors likely to have on the population density of a named plant or animal?

41 How would you demonstrate the following experimentally?
  (a) the effectiveness of a hormonal weed-killer on the weed flora of a lawn
  (b) the productivity of a standing crop, such as grass
  (c) the effect of decomposers in the soil on the rate of leaf decomposition.

10 Genetics

Genes, chromosomes, mutations, patterns of inheritance

1 In animals, which of the following corresponds most closely to the three megaspores that degenerate following meiosis in a flowering plant?
- (a) spermatozoa
- (b) polar bodies
- (c) spermatogonia
- (d) spermatocytes
- (e) spermatids.

2 If the diploid number of human chromosomes is 46, how many chromosomes would you expect to find in a spermatogonium?
- (a) 12
- (b) 23
- (c) 46
- (d) 92
- (e) 22.

3 Polysomy is the condition of having
- (a) one set of chromosomes
- (b) more than one set of chromosomes
- (c) one chromosome in the polyploid state
- (d) more than two sets of chromosomes
- (e) one or more chromosomes in the polyploid state.

4 Giant chromosomes, from the salivary glands of dipteran insects, are described as
- (a) polytene
- (b) pachytene
- (c) diplotene
- (d) leptotene
- (e) megasporocytes.

5 The term genome refers to
- (a) the total number of genes in an organism
- (b) the haploid number of chromosomes
- (c) the diploid number of chromosomes
- (d) the frequency of recessive genes in the population
- (e) the total frequency of genes in the population.

6 Chromosomes in a genome may be distinguished from one another by the position of the centromere. A chromosome with a median centromere is
- (a) eccentric
- (b) acrocentric
- (c) heliocentric
- (d) telocentric
- (e) metacentric.

7 Down's syndrome, or mongolism, is caused by an additional chromosome (no. 23) in the karyotype. This is an example of
- (a) euploidy
- (b) aneuploidy
- (c) autopolyploidy
- (d) alloploidy
- (e) pleiotropism.

8 A point mutation involves
- (a) fusion between chromosomes
- (b) a change in the base sequence of a DNA molecule
- (c) a doubling of the chromosome number
- (d) loss of a segment of a chromosome
- (e) a structural alteration in a pair of homologous chromosomes.

9 In the parental cross $AaBb \times AABb$, how many different gametes will be formed?
- (a) 1
- (b) 2
- (c) 4
- (d) 6
- (e) 8.

10 When a plant of genotype aa pollinates a plant of genotype AA, what genotype of the embryo and endosperm would be expected in the resulting seed?
- (a) Aa, Aa
- (b) Aa, AAa
- (c) Aa, AA
- (d) aa, AA
- (e) aaA, aa.

11 If a house mouse has 40 chromosomes in each somatic cell, how many autosomes are present in each gamete?
- (a) 40
- (b) 39
- (c) 20
- (d) 19
- (e) 1.

12 How many pollen grains could be produced from 40 microspore mother cells?
   (a) 160
   (b) 80
   (c) 40
   (d) 20
   (e) 10.

13 The ABO blood system of man is a classical example of
   (a) co-dominant alleles
   (b) heterosis
   (c) multiple alleles
   (d) pleiotropism
   (e) polydactyly.

14 When dihybrid plants of *Capsella* were inter-pollinated, 94% of the progeny possessed triangular seed capsules and 6% ovoid seed capsules. What is the genotype of a plant with ovoid seed capsules?
   (a) *AaBb*
   (b) *AaBB*
   (c) *Aabb*
   (d) *aabb*
   (e) *aaBB*.

15 Sickle-cell anaemia is a classical example of
   (a) a point mutation
   (b) a deletion
   (c) an insertion
   (d) a duplication
   (e) polyploidy.

16 Genes *A* and *B* are ten map units apart. Gene *C* is 5 units from *A* and 7 units from *D*, which is 8 units from *B*. What is the linear order of genes on the chromosome?
   (a) *A B C D*
   (b) *C A D E*
   (c) *B D A C*
   (d) *B C A D*
   (e) *D A B C*.

17 When a flower with white petals was crossed with a flower with red petals, the F_1 generation had pink petals. What ratio of phenotypes could be expected in the F_2 generation?
   (a) all white
   (b) all red
   (c) 50% white : 50% red
   (d) 25% white : 25% pink : 50% red
   (e) 25% white : 50% pink : 25% red.

18 Two plants, which are heterozygous for two genes, *A* and *B* are crossed. What percentage of the F_1 generation of this cross, if self-pollinated, would breed true to type?
   (a) 12.5%
   (b) 25%
   (c) 50%
   (d) 75%
   (e) 100%.

19 A mammal described as homogametic is most likely to possess
   (a) two *X* chromosomes
   (b) two pairs of homologous chromosomes
   (c) two *Y* chromosomes
   (d) an *X* and a *Y* chromosome
   (e) two sex chromosomes of the same type.

20 In mammals the sperm carries the sex chromosome(s)
   (a) *YY*
   (b) *XX*   (d) *X*
   (c) *XY*   (e) *X* or *Y*.

Short Structured Questions

21 (a) Complete the following account of enzyme synthesis in bacterial cells by supplying (i)–(vii).
   Jacob and Monod developed the hypothesis of the operon to explain how cells regulate enzyme biosynthesis. An operon is a group of three different types of related genes all aligned along a single segment of DNA. The . . . (i) gene is the site at which formation of mRNA begins; the . . . (ii) is the site of regulation, and one or more . . . (iii) genes codes for enzymes or other proteins. In some cases a fourth gene, the . . . (iv) codes for a protein, the . . . (v), which apparently binds to the site of regulation, inhibiting the action of this gene. Substances such as lactose, that increase the amount of enzyme produced by a cell, are known as . . . (vi) and the enzymes they influence are known as . . . (vii) enzymes.
   (b) Name one other compound, apart from lactose, that increases the production of a named enzyme.
   (c) Siamese cats have dark brown tips to their ears, the tip of their tail and often their paws. How do you account for this?

22 What do geneticists understand by each of the following terms?
   (a) supergene
   (b) mutation
   (c) telocentric chromosome
   (d) karyotype
   (e) epistasis
   (f) apomixis
   (g) pleiotrophy
   (h) position effect
   (i) aneuploidy
   (j) bivalent
   (k) euploidy
   (l) parthenogenesis
   (m) non-disjunction
   (n) probability.

23 Fig. 76 illustrates the process of cross-fertilisation about to take place in wheat (*Triticum vulgare*). Reproductive nuclei are given genotype letters *A* or *a*, whilst others which are vegetative are lettered differently.

(a) Which nuclei normally fuse before fertilisation takes place?

(b) Name the nuclei (i) *BCD*, (ii) *EF* and (iii) *X*.

(c) What combinations of nuclei might occur in the embryo region of the seed?

(d) What combinations of nuclei might occur in the endosperm region of the seed?

(e) If the flower had been self-fertilised, what is the most likely combination of nuclei to occur in (i) the embryo and (ii) the endosperm?

(f) What is the chief (i) advantage and (ii) disadvantage to an organism of self-fertilisation?

(g) In peas the gene for tallness (T) is dominant to the gene for dwarfness (t). If two heterozygous pea plants ($Tt \times Tt$) and their progeny are self-pollinated, what will be the percentage of homozygotes produced in the (i) F_1, (ii) F_2 and (iii) F_3 generations?

(h) Asparagus is a dioecious plant in which maleness (staminate flowers) is governed by a dominant gene P and femaleness (pistillate flowers) by its recessive allele p.

What sex ratio might be expected amongst plants resulting from cross-pollination (i) $PP \times pp$ and (ii) $Pp \times pp$?

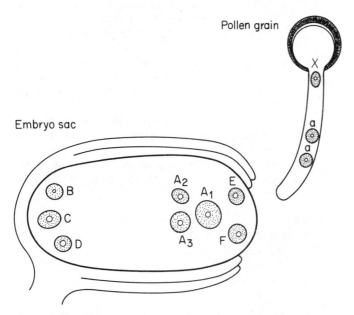

Fig. 76 The embryo sac and pollen grain of wheat immediately before fertilisation

24 Fig. 77 illustrates nuclear behaviour in a developing pollen grain and ovule of a flowering plant.

(a) Name the nuclei labelled A to F in the diagram.

(b) A pollen mother cell has the genotype *AaBb*. What will be the genotype(s) of pollen grains produced from this mother cell?

(c) An embryo sac mother cell has the genotype *AABB*. What will be the genotype(s) of its ova?

(d) If the genotype of the female parent is *Aa* and that of the male parent *Bb*, state all of the possible genotypes of the following :

Mammals	Flowering plants
(i) egg	(iv) pollen tube nucleus
(ii) sperm	(v) male gamete
(iii) foetus.	(vi) ovum
	(vii) zygote
	(viii) primary endosperm nucleus
	(ix) secondary endosperm nucleus.

(e) A sporophyte (2n) plant of a fern has the genotype *AaBbCc*. What genotypes would you expect to find amongst the prothalli (n) that grew from its spores?

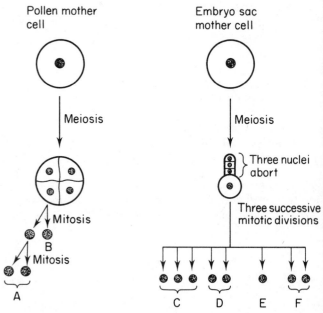

Fig. 77 Nuclear behaviour in a developing pollen grain and ovule of a flowering plant

25 Primrose flowers are either 'pin eyed' or 'thrum eyed'. Pin flowers have a long style and short anthers, lying in the corolla tube, while thrum flowers have a short style and anthers at the top of the corolla tube. Pin flowers have the genotype *ss*, while thrum eyed are heterozygous, *Ss*. Fertilisation can occur only between flowers of different genotypes.

(a) (i) How is pollination prevented in this flower?
(ii) What offspring phenotypes result from crosses between the following flowers?

Pollen		Ovule
Ss	×	*ss*
ss	×	*Ss*

Self-pollination in plants such as cherry is prevented by incompatibility between certain pollen grains and certain ovules. The controlling gene S exists as a large series of different alleles, which are designated S_1, S_2, S_3, S_4 etc. Ovules produced by a plant have two of these alleles. Pollen grains bearing either of these alleles fail to germinate on the stigma, whereas pollen grains containing other alleles grow down the style and effect fertilisation. Compatibility and incompatibility are illustrated in Fig. 78.

Pollinator : $S_1 S_2$ Pollinator : $S_3 S_4$
Pollinated : $S_1 S_2$ Pollinated : $S_1 S_2$

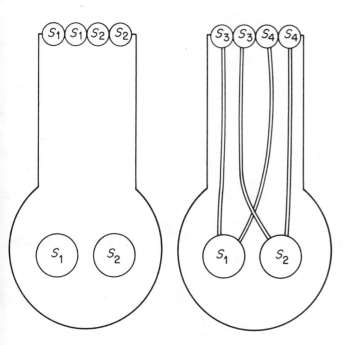

No fertilisation – pollen grains fail to grow

Zygotes : $S_1 S_3$, $S_1 S_4$, $S_2 S_3$, $S_2 S_4$

Fig. 78 Compatibility and incompatibility between pollen grains and ovules in cherry

(b) List the genotypic ratio of progeny expected from each of the following crosses:

	Pollinator	Pollinated
(i)	$S_1 S_1$	$S_1 S_2$
(ii)	$S_2 S_3$	$S_3 S_4$
(iii)	$S_1 S_2$	$S_3 S_4$
(iv)	$S_2 S_3$	$S_3 S_4$
(v)	$S_2 S_3$	$S_2 S_3$

(c) How much of the pollen is incompatible in each of the above crosses?

26 (a) Where, in a cell nucleus, are linked genes located?

(b) Why do linked genes not obey Mendel's Law of Independent Assortment?

(c) How would you test a diploid organism, with two allelomorphic pairs of genes, for linkage between the two genes?

(d) Fig. 79 shows synapsis and crossing over between pairs of chromatids of a bivalent during meiosis. Complete (i)–(iv) by showing the appearance of the chromosomes at the end of meiosis I and at the end of meiosis II.

(e) Define the term *crossover* value.

Synapsis and crossing over

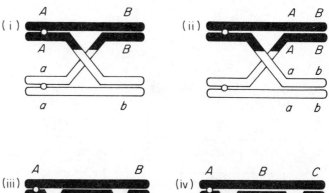

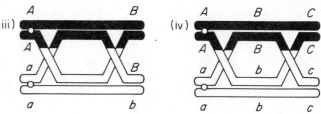

Fig. 79 Synapsis and crossing over within a pair of homologous chromosomes

(f) Five genes, A, B, C, D and E are carried on the same chromosome. The crossover value between different genes is given in Table 40. Draw a chromosome map, plotting the genes in their correct sequence, with the correct spacings between them.

(g) What is the minimum information that needed to be given in order to plot the correct sequence and position of the five genes, A, B, C, D and E?

Table 40

Genes	Crossover value
AB	20
AE	4
BC	4
BE	16
CA	16
CE	12
DA	12
DB	8

27 Comment on the table of pedigree shown in Fig. 80. Show, by using the symbols *AA*, *Aa* or *aa*, how particular genotypes may have been inherited through each pedigree.

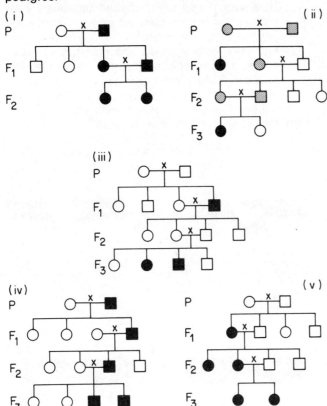

Fig. 80   Tables of pedigree

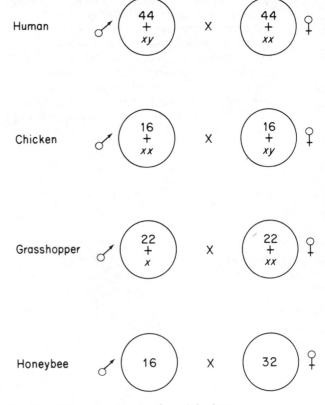

Fig. 81   Different methods of sex inheritance

28 (a) In mammals and insects sex is inherited in a number of different ways. Copy and complete Fig. 81 to show how sex is inherited in the F_1 generation of each organism.

(b) Queen bees are diploid (2n = 32) and the drones are haploid (n = 16). Queen bees lay either unfertilised eggs (n = 16), which develop into drones, or fertilised eggs (2n = 32), which develop into queens or workers.

In bees, the gene for yellow colour (*B*) is dominant to the gene for black (*b*). What would be the colour of worker and drone bees produced from the following matings?
(i) Black queen × yellow drone.
(ii) Yellow queen × black drone.

(c) It is possible for abnormalities in the distribution of chromosomes to occur during meiosis, as indicated in Fig. 82.
(i) What failure in the process of meiosis accounts for the production of these abnormal gametes?
(ii) What term could be used to describe the chromosome number in any zygote formed between one of these gametes and a normal gamete?

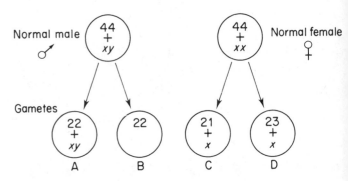

Fig. 82   Abnormalities in the distribution of chromosomes

(d) What conditions would result from fusions between a normal gamete and
(i) gamete A
(ii) gamete B
(iii) gamete D, assuming that this gamete contained an additional chromosome no. 21?

29 (a) In a certain plant the gene for tallness is dominant to the gene for dwarfness. List all of the parental genotypes that could have been crossed to give progeny that were
(i) all tall
(ii) 5 tall : 5 dwarf
(iii) 5 tall : 15 dwarf
(iv) all dwarf.

(b) The same plant has a second gene for red flowers, (*R*), which is dominant to the gene for white flowers (*r*). Draw a Punnett square to show the offspring genotypes of a cross between two plants heterozygous for both genes (*TtRr* × *TtRr*). List the offspring phenotypes resulting from this cross.

(c) What would be the phenotypic ratios if
  (i) genotype *Tt* produced individuals of intermediate height
  (ii) genotype *Rr* produced pink flowers
  (iii) genotype *Tt* produced individuals of intermediate height, and genotype *Rr* produced plants with pink flowers?

(d) How is it possible, from a cross between individuals, heterozygous for two genes, to produce offspring phenotypes in the ratio:
  (i) 9 : 7
  (ii) 15 : 1
  (iii) 12 : 4?

30 (a) In potatoes oval tubers (gene *O*) are dominant to round tubers (gene *o*) and resistance to blight fungus (gene *R*) dominant to susceptibility (gene *r*). What will be the phenotypes of the F$_1$ generation if plants are crossed which are heterozygous for both alleles?

(b) In potato blight fungus the gene for avirulence (gene *V*) is dominant to the gene for virulence (gene *v*). *R*:*V* interactions are summarised in Table 41.

Table 41

		Host		
Genotype		*RR*	*Rr*	*rr*
Parasite	*VV*	−	−	+
	Vv	−	−	+
	vv	+	+	+

A minus (−) disease reaction denotes plant resistance and a plus (+) reaction denotes susceptibility.

By reference to the F$_1$ generation of potato plants produced from homozygous parents, calculate the approximate percentage of individuals susceptible to attack from blight fungus with the genotypes (i) *VV*, (ii) *Vv* and (iii) *vv*.

31 In guinea pigs black fur (*B*) is dominant to white fur (*b*) and short hair (*S*) is dominant to long hair (*s*). If both genes obey Mendel's laws, attempt a full explanation of the following.

(a) Several matings between a white, short-haired female and a black, short-haired male produced the following progeny in the F$_1$ generation:
  6 black, short-haired

2 black, long-haired
2 white, long-haired
6 white, short-haired.

(b) A mating between a white long-haired female and the same black, short-haired male produced:
  2 black, short-haired
  2 black, long-haired
  2 white, long-haired
  2 white, short-haired.

32 In the ABO blood system of man, gene *A* and gene *B* are co-dominant, while gene *a* is recessive to both genes.

An individual has loci for any two of these genes, and the resulting phenotypes are designated groups A, B, AB and O. Group A, for example, has a genotype of *AA* or A*a*, and the genotype of group O is *aa*.

When blood transfusions are given, group O can act as the universal donor, whose red cells may be transfused into the body of any other genotype, without causing agglutination.

Individuals of blood group O, however, can receive transfusions from a Group O donor only.

Blood cells of group AB cannot be given to individuals possessing any other genotype, although AB individuals may receive blood of any type. Blood cells of group A cannot be given to recipients of group B, nor can blood cells of group B be given to recipients of group A.

(a) List all of the F$_1$ genotypes that could result from matings between the following groups:
  (i) Group A × Group B
  (ii) Group A × Group O
  (iii) Group B × Group O
  (iv) Group B × Group B
  (v) Group AB × Group O.

(b) From which of the above matings could *both* parents act as donors to all of their children?

(c) From which of the above matings could *one* parent only act as a donor to all of their children?

(d) From which of the above matings could *neither* parent act as a donor to any of their children?

33 The colour of wheat grains is determined by the combined effects of two genes, R$_1$ and R$_2$. A cross between two plants, heterozygous for both genes, produced 15 red-grained and one white-grained plant. Red-grained plants, however, could further be sub-divided into: 1 dark red: 4 medium-dark red: 6 medium red: 4 light red.

(a) Give a full explanation of genetic control of grain colour in wheat.

(b) Which plants of the F$_1$ generation, if self-pollinated, would breed true to type?

(c) List all the possible genotypes of parental plants producing F$_1$ progeny in the following proportions:
  (i) all medium red
  (ii) 1 light red: 1 white
  (iii) 1 medium red: 2 light red: 1 white
  (iv) 1 medium red: 1 light red.

34 (a) An X-linked recessive gene (*a*) produces red–green colour blindness in man. A normal woman whose father was colour-blind marries a colour blind man.

(i) What genotypes are possible for the parents of the colour-blind man?

(ii) Of the children from these parents, what percentage of the boys will be colour-blind?

(iii) What percentage of the girls will be colour-blind?

(b) A woman is a carrier of the sex-linked gene for haemophilia.

(i) What percentage of her sons will be affected if she marries a man who is normal?

(ii) Why is it extremely rare for a woman to show symptoms of haemophilia?

(c) Female cats may be black, tortoiseshell or yellow, but male cats are either black or yellow. If the colours are governed by a sex-linked locus, how can these results be explained?

35 In mammals, the female is the homogametic sex and the male is heterogametic. The X chromosome of mice carries a gene for black coat colour, (*B*), which is dominant to its allele for white coat colour (*b*). An autosome carries a gene for tail length. Gene *T*, for long-tailed, is dominant over its allele (*t*) for short-tailed.

What offspring phenotypes are likely to be obtained from the following crosses?

	Male		Female
(i)	*BTt*	×	*bbTt*
(ii)	*btt*	×	*BbTT*
(iii)	*bTt*	×	*BbTt*

36 Inheritance of flower length in tobacco is quantitative. The length of individual flowers is determined by three genes, *A*, *B* and *C*, which in dominant form each add 9 mm to a minimal length of 39 mm. Hence, in the shortest flowers

$$\left.\begin{array}{l} a \to 0 \\ a \to 0 \\ b \to 0 \\ b \to 0 \\ c \to 0 \\ c \to 0 \end{array}\right\} = 0$$

Minimal length = 39 + 0 = 39 mm.
and in the longest flowers

$$\left.\begin{array}{l} A \to 9 \\ A \to 9 \\ B \to 9 \\ B \to 9 \\ C \to 9 \\ C \to 9 \end{array}\right\} = 54$$

Minimal length = 39 + 54 = 93 mm.

Calculate the length of flowers in crosses between tobacco plants with the following genotypes:

(i) *AABB × BBCC*

(ii) *AaBb × AaBb*

(iii) *abCC × abbC*

(iv) *aabb × aabb*.

37 (a) In *Drosophila* red eye (*R*) is dominant to purple eye (*r*) and grey body (*W*) is dominant to black (*w*). A cross between a male purple-eyed, black-bodied fly and a female heterozygous red-eyed, grey-bodied fly produced offspring in the following numbers:

116 Purple-eyed, black-bodied
3 Purple-eyed, grey-bodied
5 Red-eyed, black-bodied
114 Red-eyed, grey-bodied.

What may be deduced from these results?

(b) The distance between 7 loci on a chromosome are presented in Table 42. Construct a genetic map to include these 7 loci.

Table 42

	A	B	C	D	E	F	G
A	–	4.6	16.5	10.2	.24.0	5.0	3.4
B		–	21.1	14.6	28.6	9.6	8.0
C			–	6.3	7.5	11.5	13.1
D				–	13.8	5.2	6.8
E					–	19.0	22.6
F						–	1.6
G							–

Long Structured Questions

38 What is a gene?
Describe, by reference to named examples, the different types of gene that occur on the chromosomes of eukaryotic organisms. How may (a) the structure and functions of a gene, and (b) the phenotypic expression of a gene, be influenced by environmental factors?

39 Write an illustrated account of the structure, variety, functions and behaviour of chromosomes.

40 Describe, with details of the practical techniques involved, how it is possible to map the position of genes on chromosomes.

41 Outline the experimental evidence that has led to the following conclusions.

(a) Genes are borne in the nucleus.

(b) Genes are carried on chromosomes.

(c) DNA is the material of genes.

(d) Mutations may involve faulty replication of DNA.

11 The Hardy–Weinberg Law, Evolution and Classification

Population genetics, Darwinism, neo-Darwinism, systems of classification

Multiple Choice Questions

1 The Hardy–Weinberg Law is expressed by the equation $p^2 + 2pq + q^2 = 1$. The frequency of homozygous dominant individuals in the population is therefore
- (a) p^2
- (b) q^2
- (c) $q^2 - 1$
- (d) $2pq$
- (e) $1 - q$.

2 The equation $p^2 + 2pq + q^2 = 1$ expresses the frequency of genotypes in a gene pool. If the genotype frequency of a population is $AA = 0.1$, $Aa = 0.2$ and $aa = 0.7$, the gene frequency is
- (a) $p = 0.1$, $q = 0.9$
- (b) $p = 0.2$, $q = 0.8$
- (c) $p = 0.3$, $q = 0.7$
- (d) $p = 0.4$, $q = 0.6$
- (e) $p = 0.5$, $q = 0.5$.

3 In the parent generation the frequency of gene $r = 0.8$ and that of gene $R = 0.2$. The genotype frequency of the F_1 generation will therefore be
- (a) $RR = 0.16$, $Rr = 0.64$, $rr = 0.20$
- (b) $RR = 0.10$, $Rr = 0.80$, $rr = 0.10$
- (c) $RR = 0.40$, $Rr = 0.32$, $rr = 0.64$
- (d) $RR = 0.60$, $Rr = 0.30$, $rr = 0.10$
- (e) $RR = 0.04$, $Rr = 0.32$, $rr = 0.64$.

4 The first living organisms, having arisen in a sea of organic molecules, are believed to have fed as
- (a) autotrophs
- (b) heterotrophs
- (c) saprophytes
- (d) carnivores
- (e) herbivores.

5 With whom do you associate the view that 'structures are acquired, enlarged, reduced or lost through use and disuse, and these changes are inherited by subsequent generations'?
- (a) Darwin
- (b) Lamarck
- (c) Mendel
- (d) Wallace
- (e) Oparin.

6 Darwin's *Origin of Species* was published in
- (a) 1809
- (b) 1838
- (c) 1848
- (d) 1859
- (e) 1881.

7 Which of the following was the greatest weakness in Darwin's original theory of evolution by natural selection?
- (a) poor fossil evidence
- (b) failure to appreciate the nature of inheritance
- (c) failure to understand the nature of selection
- (d) failure to identify the forces of selection
- (e) failure to assemble sufficient evidence of evolution in plants.

8 The limbs of an insect and the leg of a bird are
- (a) analogous
- (b) homologous
- (c) degenerate
- (d) analogous + homologous
- (e) neither analogous nor homologous.

9 Adaptive radiation describes
- (a) the colonisation of a new area by the descendants of an original pair of colonisers
- (b) the origin of new organs
- (c) the evolution of island faunas
- (d) niche selection
- (e) the evolution from a single ancestral species of a variety of species, each adapted to different habitats.

10 Cases in which the developmental system has been altered in evolution so that an intermediate ancestral growth stage becomes the terminal form in the descendant is known as
- (a) epistasis
- (b) neoteny
- (c) mutation
- (d) parthenogenesis
- (e) autogamy.

11 If variations among groups of organisms follow a geographical distribution related to temperature, humidity and other environmental conditions, such graded variation is known as a
- (a) cline
- (b) clone
- (c) sere
- (d) niche
- (e) zone.

12 Which of the following is most likely to become fossilised?
  (a) *Amoeba*
  (b) a green alga
  (c) a herbaceous plant
  (d) an insect
  (e) a vertebrate.

13 Fossils more than 40 000 years old are generally dated by reference to the decay of
  (a) carbon-14
  (b) nitrogen-15
  (c) oxygen-18
  (d) uranium-238 → lead-206
  (e) caesium-137

14 Variation in the peppered moth (*Biston betularia*) affords an example of
  (a) autopolyploidy
  (b) balanced polymorphism
  (c) transient polymorphism
  (d) aneuploidy
  (e) epistasis.

15 Which of the following organisms is classified as a primitive eukaryote?
  (a) virus
  (b) bacterium
  (c) blue-green alga
  (d) green alga
  (e) mushroom.

16 Which of the following groups is listed in descending order of hierarchy?
  (a) kingdom → class → phylum → order
  (b) kingdom → order → phylum → class
  (c) kingdom → class → genus → species
  (d) class → order → genus → species
  (e) order → genus → family → species

17 Which of the following features is *not* shared by the arachnids and insects?
  (a) bilateral symmetry
  (b) a ventral nerve cord with paired ganglia
  (c) one pair of antennae
  (d) a hard jointed exoskeleton
  (e) paired jointed appendages.

18 Which group of organisms is described as eukaryotic and heterotrophic, with nuclei occurring in a basically continuous mycelium?
  (a) bacteria
  (b) fungi
  (c) blue-green algae
  (d) algae
  (e) bryophytes.

19 If a new plant were to be discovered with parallel-veined leaves, and brightly coloured flowers each with 6 stamens, the plant would most probably be classified as

  (a) a gymnosperm
  (b) an angiosperm
  (c) a monocotyledon
  (d) a dicotyledon
  (e) a pteridophyte.

Short Structured Questions

20 The frequency of each genotype in a randomly breeding population can be found by using the Hardy–Weinberg equation:

$$p^2 + 2pq + q^2 = 1$$

In this equation p = frequency of gene A, and q = frequency of gene a.
  (a) The frequency of genes A and a in six different populations is given in Table 43. Calculate, for each population, the genotypic frequencies of AA, Aa and aa, Plot these results as a graph.

Table 43

Population	Frequency gene A	Frequency gene a
1	1.0	0.0
2	0.8	0.2
3	0.6	0.4
4	0.4	0.6
5	0.2	0.8
6	0.0	1.0

  (b) Assuming you could recognise the phenotype of recessive individuals, of what practical value would the graph be in studying a population of 100 laboratory mice, the product of random matings, without selection?
  (c) A farmer has 100 male sheep (90 black, 10 white) and 100 female sheep (70 black, 30 white). Assuming that mating occurs randomly, what are the expected frequencies of the various matings?
  (d) Calculate the expected genotype frequencies in the next generation in each of the following populations:
    (i) AA = 0.64, Aa = 0.32, aa = 0.04
    (ii) AA = 0.70, Aa = 0.20, aa = 0.10
    (iii) AA = 0.40, Aa = 0.40, aa = 0.20
    (iv) AA = 0.66, Aa = 0.32, aa = 0.02
    (v) Which of the above populations is in Hardy–Weinberg equilibrium?
  (e) List the factors that may cause changes in the gene frequency of a population.
  (f) A recessive gene occurs in one person in 20 000. What percentage of the population is heterozygous for this gene?

21 Apply the Hardy–Weinberg Law to each of the following problems:
  (a) White plumage is dominant to black plumage. A

population of 900 chickens consists of 891 white and 9 black birds. If the white birds have the genotypes *BB* and *Bb*, estimate the allelic frequencies of *B* and *b* in the population.

(b) The ability of certain people to taste a chemical compound called PTC is governed by a dominant allele *T*, and the inability to taste PTC by its recessive allele *t*. If the population is composed of 24% *TT*, 40% *Tt* and 36% *tt*, what is the frequency of genes *T* and *t*?

(c) It has been suggested that European eels and American eels are derived from a single spawning population that breeds annually in the Sargasso Sea. Analysis of haemoglobin types in the two adult populations, however, gave the results shown in Table 44. How would you interpret these results?

Table 44

Population	% genotypes in population		
	AA	*Aa*	*aa*
American eels	83	14	3
European eels	100	0	0

22 Name *two* important examples of each of the following:

(a) evolutionary theorists
(b) forces of natural selection
(c) populations of animals which are polymorphic
(d) co-evolution
(e) adaptive radiation
(f) extinct species
(g) marsupial mammals
(h) 'living fossils'
(i) hybrids
(j) 'missing links'
(k) convergent evolution
(l) homologous structures
(m) vestigial structures.

23 (a) Measurements were made of ear length and tail length in a population of mice living in the wild. Results are illustrated in Fig. 83.
What type of variation is shown by (i) ear length and (ii) tail length?

(b) Assuming that natural selection has not acted on these populations, what information about the nature of the genes controlling (i) ear length and (ii) tail length may be obtained from the graphs?

(c) To which of these characters is the Hardy–Weinberg Law applicable?

(d) What additional information would be required before the Hardy–Weinberg Law could be applied?

(e) Populations of organisms living in the wild are subject to three different types of natural selection, as indicated in Fig. 84. State briefly the effect of each type of selection on any population.

(f) What would happen, in time, if each of these types of selection had acted on the population of mice with respect to ear length?

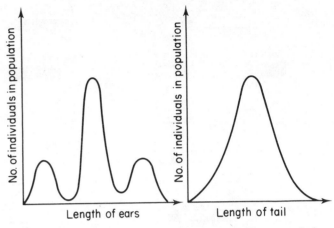

Fig. 83 Variations in ear-length and tail-length in a population of mice

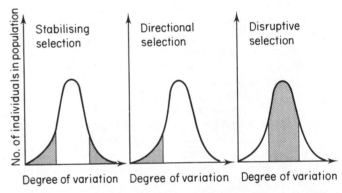

Fig. 84 The effects of different types of selection on populations. Shaded portions indicate parts of populations selected against

(g) By reference to a population of field mice, name *five* different forces of selection, in addition to predation, that could exert a selective effect on the population during the course of a year.

(h) Populations of wood mice often make their nests in healthland, burrowing into sandy soils, up to 30 cm in depth, which overlie sandstone. A study of these nests showed that most were built at depths of between 15–20 cm below the surface. There was some evidence to suggest that those mice which built either deeper in the soil, or more superficially, were less successful in rearing their young. Bearing in mind that the woodmouse has a number of predators, that water-retaining iron pans are a feature of healthland soils, and that fires frequently destroy surface vegetation in the spring, attempt to define those forces of natural selection which regulate the depth of nesting in the soil.

24 (a) What do you understand by a *mutation*?
   (b) Fig. 85 shows changes that can occur in the linear arrangement of genes on chromosomes. What name is given to each of these mutations?

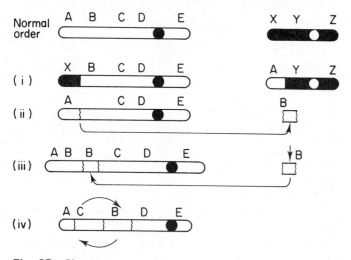

Fig. 85 Chromosome mutations

   (c) Name types of mutation that involve
      (i) the entire genome of an organism
      (ii) a DNA codon
   (d) Give, in each case, *three* examples of mutagenic agents that are
      (i) chemical compounds
      (ii) types of radiation.
   (e) If a mutation causes a change in the base sequence of a DNA molecule, why does this not invariably lead to the synthesis of a different polypeptide by incorporation of a foreign amino acid?

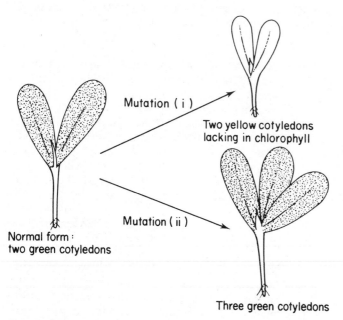

Fig. 86 Mutations in seedlings of sunflower

   (f) When a gene mutates,
      (i) does the mutation affect one or both alleles?
      (ii) is the resulting individual heterozygous or homozygous for the mutated gene?
   (g) Is it possible for a gene to mutate to (i) a recessive or dominant form and, if so, (ii) which is the most frequent?
   (h) Fig. 86 shows two mutations that may occur in seedlings of sunflower. Comment on the detrimental or beneficial effects that these two mutations might confer on those seedlings in which they occur.

25 Fig. 87 shows the development of yeast colonies on a plate of nutrient agar. At the same time that the plate was seeded with yeast cells, a solution of acriflavine was introduced into the central well. The plate was incubated at 30° C for 24 hours. At the end of incubation it appeared as shown in the diagram.
   (a) Attempt to provide an explanation of the zonation of colonies around the central well containing acriflavine.

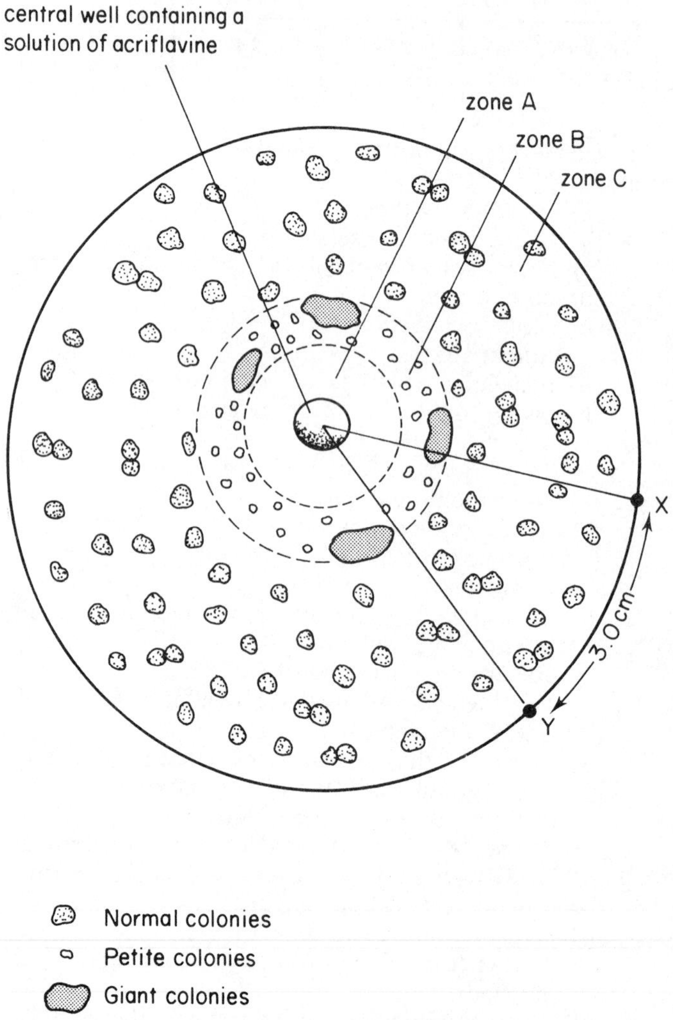

central well containing a solution of acriflavine

zone A
zone B
zone C

X

3.0 cm

Y

⊛ Normal colonies

○ Petite colonies

⬭ Giant colonies

Fig. 87 The effect of acriflavine on the development of yeast colonies on an agar plate

(b) What terms are applied to substances which exert an effect on living cells similar to the effects of acriflavine on yeast?

(c) If the seeded plate of yeast had been incubated at 37° C instead of 30° C, what effects might this have had on the appearance of the plate?

(d) Count the number of normal and petite colonies in the sector of the dish labelled X–Y. If the dish has a radius of 4.5 cm and the distance X–Y is 3.0 cm., calculate the approximate number of normal and petite colonies in the dish. Show how you arrive at your answer.

26 Darwin and the neo-Darwinists view the origin of a new species from different backgrounds and in different terms. Essential concepts in the Darwinian and neo-Darwinian hypotheses are set out in Fig. 88.

Give concise definitions of each of the terms (i)–(xi) used in the diagram, making clear essential differences between the Darwinian and neo-Darwinian hypotheses.

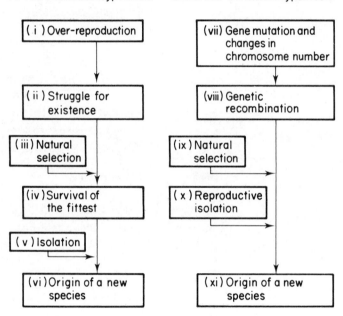

A. The Darwinian Hypothesis   B. The neo-Darwinian Hypothesis

(i) Over-reproduction

(ii) Struggle for existence

(iii) Natural selection

(iv) Survival of the fittest

(v) Isolation

(vi) Origin of a new species

(vii) Gene mutation and changes in chromosome number

(viii) Genetic recombination

(ix) Natural selection

(x) Reproductive isolation

(xi) Origin of a new species

Fig. 88 Essential concepts in Darwinian and neo-Darwinian views of the origin of new species

27 Fig. 89 illustrates two models of the evolutionary process.

(a) Comment on the chief differences between these two models.

(b) When Darwin first advanced his theory of evolution, what aspects of the process were illustrated by reference to
(i) artificial selection
(ii) the fauna of the Galapagos Islands
(iii) Darwin's finches?

(c) Populations of the land snail *Cepaea nemoralis* are polymorphic. The basic colour of the shell may vary from yellow, through green, to brown.

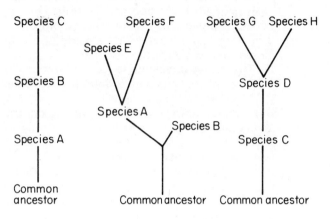

Fig. 89   Models of evolution

In addition, the snail may be unbanded, or banded with from 1–5 dark bands around the shell. Analysis of the morphs from three different localities is given in Table 45.

Table 45

Shell pattern/ colour	% each type in population		
	Woodland	Sward	Grassland
Green-yellow background	10	16	63
Banded	32	48	28
Dark brown background	58	36	9

(i) What selective pressures are likely to be in operation, and how do you account for differences in the composition of populations in the different localities?

(ii) Studies of the selective pressure exerted by predatory birds on yellow-green snails from the floor of a wood revealed seasonal differences in the pattern of selection (Table 46). How would you account for these results?

(iii) It is known from studies of fossil snail shells that polymorphism has existed since before the Neolithic period. If predation alone cannot account for the prevalence of particular morphs in certain localities, what other factors might be involved in the process of selection?

Table 46

Season	No. yellow-green snails eaten by birds
Winter	51
Spring	35
Summer	11
Autumn	13

28 (a) Fig. 90 shows an outline classification, listed in order of increasing complexity of major groups of living organisms, from which some entries have been omitted. Make appropriate entries in (i)–(xv) to complete the outline classification.

(b) Which of the following statements apply to mosses and which apply to ferns?

(i) The spores germinate to form a prothallus.

(ii) The spores germinate to form a protonema.

(iii) The leaves are sporophylls.

(iv) The mature sporophyte is entirely dependent upon the gametophyte for its supply of nutrients.

(v) The sporophyte possesses true roots.

(vi) The gametophyte plant possesses unicellular rhizoids.

(vii) Antheridia and archegonia are usually borne on separate plants.

(viii) The leaves are simple, without a petiole.

(ix) The spores are liberated through peristome teeth.

(x) The spores develop within sporangia.

(c) Which of the following statements apply to amphibians and which apply to reptiles?

(i) Larvae, known as tadpoles, may be aquatic.

(ii) Fertilisation is external.

(iii) Fertilisation is internal.

(iv) Mating, fertilisation and development take place in water.

(v) Waterproof keratinised scaly skin.

(vi) The heart is completely divided into two halves.

(vii) Large heavily-yolked egg with protective shell.

(viii) The heart is three-chambered.

(ix) There is a single auditory ossicle in the middle ear.

(x) Includes salamanders and toads.

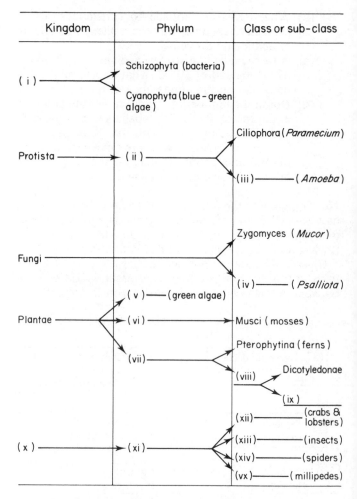

Fig. 90　Outline classification of living organisms

29 Complete the Tables 47 and 48, either by naming appropriate organisms or by listing diagnostic features.

Table 47

Comparative features	Mammals	Birds	Amphibians	Fishes
(i) Class				
(ii) Two named examples living in the wild				
(iii) Nature and functions of skin				
(iv) Fertilisation				
(v) Protection of the embryo				
(vi) Organs of respiration				

Table 48

Comparative Feature	Locust	Potato	*Mucor*
(i) Kingdom			
(ii) Phylum			
(iii) Class			
(iv) Mode of nutrition			
(v) Asexual reproduction			
(vi) Chief component of cell walls			

Long Structured Questions

30 What is the relevance of the Hardy–Weinberg Law to the study of (a) populations and (b) evolution?

31 Give an account of Darwin's theory of evolution. Of what relevance to his hypothesis was
    (a) Malthus' essay on population
    (b) artificial selection
    (c) the fauna of the Galapagos Islands?

32 Outline the evidence for evolution that was available to Darwin.
What further evidence is now available?

33 To what extent do modern concepts of evolutionary processes draw upon Darwin's work?
In what ways have genetical studies contributed towards establishing these concepts?

34 Evaluate the role of (a) natural selection and (b) isolation in determining the course of evolution. What other factors may influence the course of evolution?

35 Distinguish between
    (a) Mullerian and Batesian mimicry
    (b) microevolution and macroevolution
    (c) homology and analogy
    (d) chromosome and gene mutations
    (e) 'missing links' and 'living fossils'.

36 What are the causes and effects of mutations? Show how particular mutations may be (a) detrimental or (b) beneficial to those organisms in which they occur?

37 'The great majority of species that have ever existed are now extinct. In the past, extinction has made way for new species. Man is now the dominant force in the extinction of species. We do not yet know what species, if any, will fill the niches now being vacated.' Helena Curtis.
Illustrate and discuss this statement.

38 Can man learn any lessons from (a) the success and (b) the extinction of the dinosaurs?

39 What evidence in favour of evolution has been obtained from
    (a) fossils
    (b) the classification of organisms
    (c) the study of larval forms?

40 What principles are applied to the classification of plants and animals? Illustrate your answer by reference to a named phylum of plants and a named phylum of animals.

41 Give a full classification of each of the following organisms: *Amoeba*, an earthworm, a cabbage white butterfly, a frog, a rabbit, *Spirogyra*, *Mucor* and the male fern.

Answers

1 Macromolecules

1 (e) **2** (e) **3** (d) **4** (c) **5** (a) **6** (b)
7 (b) **8** (a) **9** (d) **10** (c) **11** (a) **12** (b)
13 (c)

14 (a) Glycogen (b) Starch (c) Cellulose (d) Sucrose (e) Maltose (f) Lactose (g) Glucose (h) Fructose (i) Ribose (j) Deoxyribose

15 (a) See Fig. 91.

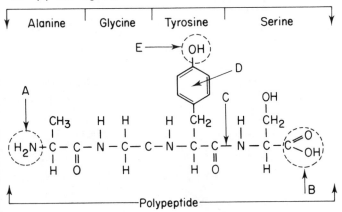

Fig. 91 A polypeptide

(b) The glycine molecule is added to the carboxyl end of the polypeptide.
(c) (i) Condensation, with release of water (ii) Hydrolysis

16 (a) Insulin: hormone
Haemoglobin: blood pigment
Myosin: contractile component of muscle
Immunoglobin: antibody
(b) Peptide bonds, disulphide bridges
(c) If biuret reagents A and B are added to a globular protein, a purple pigment is formed. Alternatively, if a protein is boiled with Millon's reagent a brick-red precipitate is obtained.

17

Glu–ala
  or   ser–tyr–leu–asp–arg–glu–ilu–his–lys–COOH
Ala–glu

18 (a) CGACGATTTAATTTTGCA
(b) GCUGCUAAAUUAAAACGU
(c) GCTGCTAAATTAAAACGT (d) (i) 18% (ii) 18%

19 (a) The surface of enzyme molecules, in the region of the active site, provides a template into which molecules are brought to positions where they can react, as a key fits a lock.

(b) Co-enzyme. NAD (nicotinamide-adenine dinucleotide), a derivative of nicotinamide, and FAD (flavin-adenine dinucleotide), a derivative of riboflavin, are both co-enzymes that act as hydrogen acceptors in reactions catalysed by dehydrogenases.
(c) Active site
(d) Salivary amylase requires Cl^- ions. Carbonic anhydrase requires Zn^{2+} ions.
(e) When the non-competitive inhibitor is bound to the allosteric site the overall shape of the molecule is changed, including the shape of its active site, which can no longer bind molecules of substrate. Organophosphorus insecticides are non-competitive inhibitors which inactivate the enzyme cholinesterase, essential for the transmission of a nervous impulse from one nerve cell to another.
(f) Competitive inhibitors are of a similar shape to substrate molecules. A competitive inhibitor attaches to the active site of the enzyme, competing for the site with substrate molecules. Malonic acid is a competitive inhibitor of succinic acid, in the conversion of succinic acid to fumaric acid, catalysed by succinic dehydrogenase.
(g) Enzyme product

20 (a) Cholinesterase: choline and acetate
(b) Thrombin: fibrin (c) Catalase: oxygen and water
(d) Trypsin (or an exopeptidase): dipeptides and amino acids
(e) Urease: carbon dioxide and ammonia

21 (a) Before the figures are plotted, as in Fig. 92, it is necessary to convert them into reciprocals of time; that is, $1/t$, or, more conveniently in this case, $(1/t) \times 10$.

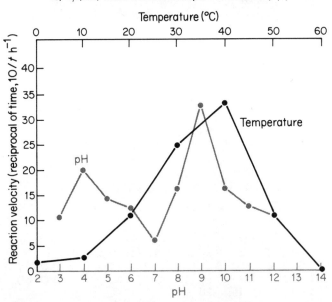

Fig. 92 Relationship between temperature, pH and the rate of reaction

(b) As temperature increases, up to an optimum of 40°C, there is a directly related increase in enzyme activity. Above 40°C, however, the rate of the reaction falls as the enzyme is denatured.

The testis of the rat contains at least one acid phosphatase and one alkaline phosphatase. Optimum pH for the acid phosphatase is 4.0 and that for the alkaline phosphatase 9.0. Other phosphatases may be present in the material.

(c) One source of inaccuracy is that the experiment investigates enzyme activity at temperature differences of 10°C and pH differences of 1.0 unit. More accurate results would have been obtained by setting up a larger number of samples at temperature differences of, say, 5°C and pH differences of 0.5 units.

22 (a) Enzymes are soluble, colloidal molecules which are produced by living cells. All are globular proteins, with a complex three-dimensional structure, capable of binding substrate molecules to a part of their surface.

(b) See Fig. 93.

(c) It is assumed that the amount of enzyme remains constant.

(d) Graph A would have differed in that optimum activity of trypsin occurs at pH 8–9. The other graphs would have been identical.

(e) The Q_{10} of an enzyme is its temperature coefficient.

$$\text{Temperature coefficient} = \frac{\text{Reaction velocity at } T+10°C}{\text{Reaction velocity at } T°C}$$

23 Vitamin B_1: b
Vitamin B_6: i
Vitamin C: a, d, e, f, j
Vitamin D: c, g, h

24 (a) Islets of Langerhans: reduces blood glucose levels

(b) Adrenalin: prepares for increased activity (fight or flight)

(c) Cortisol: adrenal cortex

(d) Posterior pituitary: promotes reabsorption of water from kidney tubules into blood

(e) Follicle stimulating harmone (FSH): anterior pituitary

(f) Thyroxin: increases basic metabolic rate (BMR)

(g) Posterior pituitary: causes contraction of uterine wall during labour

25 Gibberellin: b, d, g
Auxin: a, c, f
Ethene: e

26 (i) h   (ii) j   (iii) f   (iv) a   (v) e   (vi) i   (vii) c
(viii) b   (ix) g   (x) d

2   Cells, Organelles and Membranes

1 (c)   **2** (b)   **3** (c)   **4** (b)   **5** (a)   **6** (c)
7 (c)   **8** (e)   **9** (d)   **10** (c)   **11** (d)   **12** (c)
13 (b)   **14** (c)   **15** (a)   **16** (a)   **17** (d)
18 (b)   **19** (e)   **20** (c)

A. The effect of pH at optimum temperature

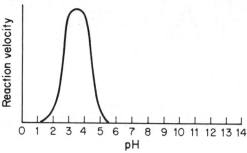

B. The effect of temperature at optimum pH

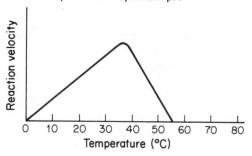

C. The effect of enzyme concentration at optimum pH and temperature

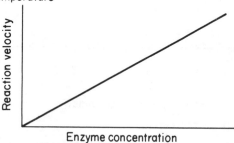

D. The effect of substrate concentration at optimum pH and temperature

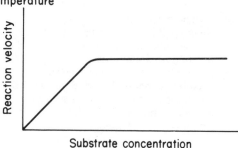

Fig. 93   Aspects of enzyme activity

21 (a) Epidermis; cortex; xylem   (b) Cortex of root; endodermoid parenchyma of stem; root cap   (c) Root hairs   (d) Xylem vessels and fibres   (e) Sieve elements   (f) Root cap; endodermoid parenchyma of stem   (g) Guard cells of stomata   (h) Xylem vessels   (i) Mesophyll cells of leaf; cortical cells of stem   (j) Apical meristems and vascular cambium   (k) Stamens and ovules of flowers

22 (a) Bone   (b) Cartilage   (c) Nervous system
(d) Heart   (e) Skin   (f) Stomach   (g) Nervous system
(h) Blood   (i) Trachea; fallopian tubes   (j) Buccal cavity; upper region of oesophagus; vagina   (k) Testis   (l) Eye
(m) Nervous system

23 (a) Chloroplasts   (b) Ribosomes   (c) Mitochondria
(d) Nucleolus   (e) Golgi bodies   (f) Centrosome
(g) Mitochondria   (h) Contractile vacuole   (i) Lysosome
(j) Plasmodesmata   (k) Vacuole   (l) Cilia or flagella

24 Mitosis: a, b, d, j
Meiosis: c, e, f, g, h, i

25 (a) A: aster   B: centrosome   C: centioles. D: centromere.   E: metaphase plate (equator).   F: spindle.
(b) Metaphase is illustrated. Prophase and prometaphase have preceded this phase; anaphase and telephase will follow it.
(c) See Figs. 94 and 95. Two drawings are required, as meiosis has metaphase I and metaphase II. The second of these metaphases occurs at right angles to the first.

Fig. 94 The appearance of the chromosomes during metaphase I. Homologous chromosomes lie in pairs at the equator, having exchanged homologous portions during meiotic prophase

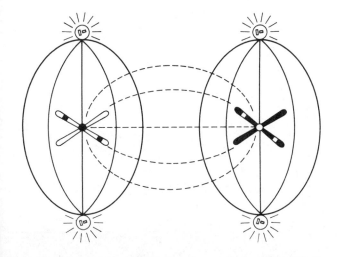

Fig. 95 The appearance of the chromosomes during metaphase II. The second metaphase is at right angles to the first and involves the formation of two spindles

26 (a) A: eyespot   B: contractile vacuole   C: basal granule (blepharoplast).   D: nucleus.   E: chloroplast.
F: myoneme.   G: pellicle.   H: flagella.   I: reservoir.
J: contractile vacuole.   K: cytostome.   L: food vacuole.   M: micronucleus.   N: macronucleus
O: cilia.
(b) (i) A and B   (ii) B   (iii) A   (IV) B   (v) A
(vi) B   (vii) A   (viii) A   (ix) B   (x) B

27 (a) X: mitochondria.   Y: chloroplast.
Both types of organelle would occur in the mesophyll cells of a leaf.
(b) A: outer membrane.   B: inner membrane.   C: DNA.
D: free ribosome.   E: crista.   F: fluid matrix.
G: starch grain.   H: grana.
I: stroma.   J: lamellae.
(c) See Fig. 96.
(d) X: i, iv, v, vi, vii, viii, x
Y: ii, iii, v, vii, ix

Organelle X   Organelle Y

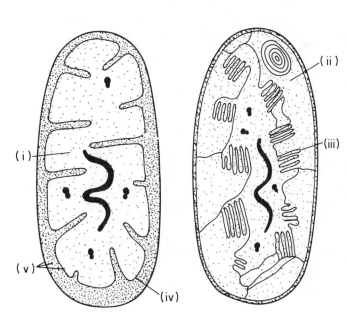

Fig. 96 Diagrammatic representation of two cell organelles

28 (a) X: Davson–Danielli model
Y: Fluid mosaic model
(b) A: proteins.   B: phospholipid.   C: cholesterol.
D: polysaccharide.
(c) In the Davson–Danielli model there is a bimolecular layer of phospholipid molecules, sandwiched between surface layers of protein. Additionally, protein molecules line pores, through which materials make their entry or exit.
    Following the identification of additional molecules in the membrane, it became necessary to construct a more complex model. Moreover, experiments on the active transport of ions showed that some molecules, notably carriers of ions, were capable of movement within the membrane. This led to the development of the fluid mosaic model in which larger molecules, suspended in a fluid matrix of phospholipid molecules are capable of occupying alternative positions.

3 Water, Solutes and Transport Systems

1 (c)   **2** (a)   **3** (a)   **4** (d)   **5** (b)   **6** (e)
7 (d)   **8** (c)   **9** (b)   **10** (c)   **11** (d)

12 (a) A: cell wall. B: plasmalemma. C: cytoplasm.
      D: tonoplast.   E: plasmodesma.   F: vacuole.

     (b) Movement of water occurs either via the apoplast (cell wall and intercellular spaces) or symplast (cytoplasm and plasmodesmata).

     (c) (i) In a molar solution of glucose, which is hypertonic to the cell sap, water would be lost from the cell by exosmosis. Firstly, the cell would become flaccid, then it would become plasmolysed, as the cytoplasm shrank. In the fully plasmolysed cell, the vacuole would be almost empty of water and the cytoplasm would remain in contact with the wall only at the plasmodesmata.
        (ii) In distilled water, which is hypotonic to the cell sap, water would enter the cell by osmosis, resulting in an increase in cell volume to a point where the cell became fully turgid.

     (d) and (e) See Fig. 97.

     (f) The term *water potential* describes the energy status of water. It may be defined as the difference in chemical potential per unit volume between a given water sample and pure water at the same temperature and pressure.

     (g) The water potential of pure water is 0.

     (h) The effect of adding solute to water is to lower the water potential. All solutions have a water potential less than 0, expressed as a negative quantity.

     (i) A fully turgid cell never contains pure water. As it contains a very dilute solution of salts, its water potential must be below 0.

     (j) Water potential relates directly to the capacity of a cell for water absorption. A flaccid cell, with a low water potential, has the greatest capacity for water absorption. A fully turgid cell, on the other hand, has little, if any, further capacity for water absorption, indicated in terms of water potential by a figure close to 0.

     (k) (i) Cell X has a water potential of -0.7 MPa and cell Y of -0.3 MPa.
        (ii) Water will move from cell Y to cell X with a force of -0.4 MPa.

13 (a) A: afferent arteriole.   B: Bowman's capsule.   C: Malpighian corpuscle.   D: proximal convoluted tubule.   E: descending limb.   F: ascending limb.   G: distal convoluted tubule.   H: collecting duct.

     (b) All the statements are correct, except statements (ii) and (v).

14 (a) See Fig. 98.

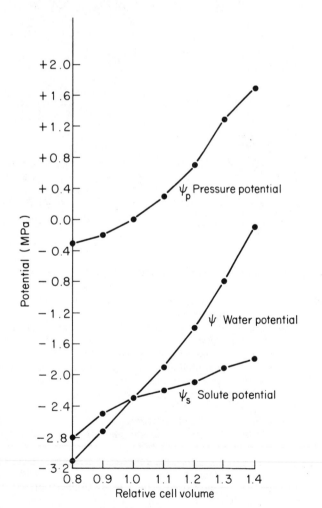

Fig. 97 The relationship between cell volume, water potential and its component potentials

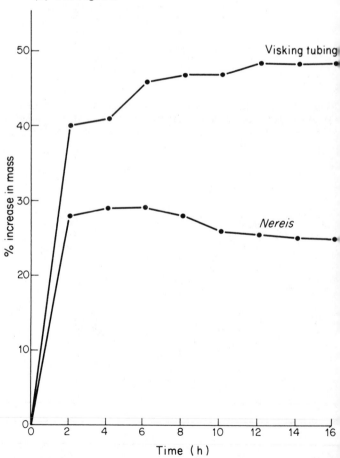

Fig. 98 Percentage increase in mass of *Nereis* and Visking tubing when immersed in 50% sea water

(b) In Visking tubing water enters through the membrane by osmosis, until the tube is fully turgid and the membrane fully stretched. The only factor that could account for water loss from the tubing is an increase in pore size, sufficient to permit the exit of solute molecules.

Nereis is capable of reducing the permeability of its body surface both to the outward diffusion of ions and the inward diffusion of water. Osmosis accounts for the initial gain in mass but, unlike the Visking tubing, *Nereis* can osmoregulate by increasing the volume of its urine. Hence, the osmotic flow of water is counteracted largely by the production of a copious urine. Additionally, *Nereis* shows an adaptation to a change in its environment, slowly reducing the salt content of its body fluids to a level that is closer to the salt content of the surrounding water. This process tends, in time, to reduce the inward osmotic flow of water.

15 (a) A molar solution is one that contains a mole (gram molecular weight) of a solute dissolved in just enough solvent to make exactly one litre (1000 cm³) of solution.

(b) (i) 29.22 g NaCl made up to 500 cm³ solution with distilled water.
(ii) 1.50 g urea made up to 250 cm³ solution with distilled water.
(iii) 2.70 g glucose made up to 150 cm³ solution with distilled water.
(iv) 4.27 g glucose made up to 25 cm³ solution with distilled water.

(c) (i) 9500 cm³ distilled water.
(ii) 250 cm³ distilled water.
(iii) 150 cm³ distilled water.
(iv) 225 cm³ distilled water.

(d) The osmotic pressure of a solution is the potential maximum pressure which could be developed across a semipermeable membrane, if the membrane were to separate the solution from pure solvent.

The osmotic pressure of all molar solutions is 22.4 atmospheres at 0° C. Hence, if the molarity of a solution is known, its osmotic pressure can be calculated. Similarly, molarities can be calculated from measurements of osmotic pressure.

(e) (i) 17.92 atmospheres (ii) 11.2 atmospheres (iii) 5.6 atmospheres.

(f) (i) There is approximately 4.88 g glucose/fructose mixture in 100 cm³ fruit juice.
(ii) Isotonic means that fluids inside and outside the cell exert an identical osmotic pressure (that is; they are identical in molarity).

16 (a) Passive diffusion is the movement of molecules from areas of high concentration to areas of low concentration.

(b) The rate of diffusion is determined by the size of the diffusing molecules, their rate of movement, the concentration gradient that exists across the membrane, and the permeability of the membrane, which is itself affected by thickness and pore size.

(c) (i) Water diffuses from the leaves and stems of green plants into the surrounding atmosphere. Similarly, water diffuses from the skin, eyes, mouth and nostrils of mammals.
(ii) Oxygen diffuses from the leaves of green plants when they photosynthesise. In mammals, oxygen dif-

fuses from the alveoli of the lungs into the blood stream.
(iii) Carbon dioxide diffuses from the organs of green plants, or from the lungs of a mammal, during respiration.
(iv) Ammonium ions, derived from the deamination of dietary proteins, diffuse from the body surface of protozoans or the gills of fishes into the surrounding water.

(d) Exchange diffusion involves an exchange of ions between an organism and its environment. H^+ and OH^- ions, produced by root hairs, can be exchanged for other cations and anions in the soil. Certain freshwater invertebrates, such as crayfish, accumulate sodium ions in exchange with ammonium and hydrogen ions, whilst chlorine is exchanged for bicarbonate ions.

(e) Osmosis is the diffusion of molecules of a solvent, such as water, through a semipermeable membrane or one that is impermeable to solutes.

(f) Osmosis plays an important role in determining the fluid content of the blood capillaries. As blood passes through the capillaries, plasma proteins are retained within the vessels, whilst water and certain solutes are filtered out through the capillary walls as a result of hydrostatic pressure. The proteins which are retained possess a relatively low water potential that causes the reabsorption of much of the water into the capillaries.

(g) Active transport is the assisted movement of ions and molecules across a membrane against a concentration gradient. This type of transport is effected by carrier molecules and dependent upon energy supplied by energy-donor substances such as ATP.

(h) Active transport probably occurs in all living cells, both across the cell surface membrane and across the membranes that surround the different organelles.

17 (a) Nitrogen, calcium, magnesium, potassium, phosphorus, sulphur

(b) Iron, manganese, boron, copper, zinc

(c) (i) Nitrogen-deficient plants have light green upper leaves, yellow mature leaves, and dry, brown older leaves. The stems may become short and slender, often tinged red by the presence of anthocyanin pigments.
(ii) Nitrogen is of extreme importance to plants because it is a constituent of proteins, nucleic acids, phytohormones etc.

(d) (i) Diffusion, leading to salt accumulation
(ii) Ion (diffusion) exchange
(iii) Active transport

(e) Ions may be absorbed via the root tip or from leaf surfaces.

(f) (i) NO_3^- and NH_4^+ (ii) SO_4^{2-}
(iii) PO_4^{3-}, HPO_4^{2-} and $H_2PO_4^-$
(iv) Fe^{2+}, Fe^{3+}, $Fe(OH)_2^+$, $FeOH^{2+}$

(g) K^+, NO_3^- and SO_4^{2-} ions are actively absorbed, whereas Na^+ and Ca^{2+} ions are actively excluded.

18 Model A: 'revolving door'
Model B: 'moving carrier'

In model A the large carrier spans the entire width of the membrane. At one side, or on both sides of the carrier are sites into which the transported ion or molecule locks. The carrier then rotates within the membrane, transporting ions etc. from one side of the membrane to the other. Sodium and potassium ions are believed to be transported across root hairs by this mechanism.

Model B shows a moving carrier, which is a relatively small molecule. Having collected ions or molecules on one side of the membrane it carries them across to the other side, where they are released. Glucose absorption from the intestine is believed to operate by this mechanism.

19   Xylem: ii, iii, vii, viii, ix, x, xi, xii
Phloem: i, ii, iv, v, vi, x, xii

20   (a) Xylem vessels transport water and dissolved mineral salts in an upward direction through a plant, from its roots to its stems and leaves. Xylem fibres, especially in stems, form rigid columns, or a rigid cylinder of mechanical tissue, that supports the plant against the force of gravity.
(b) (i) Xylem vessels formed in spring are generally larger and more numerous than those formed in the autumn.
(ii) Vessels in the 'sapwood' are open and conduct water; those in the 'heartwood' are non-conducting, having been blocked by resinous deposits called tyloses.
(c) Osmosis can occur only between cells that are separated by a semi-permeable membrane. Xylem vessels are dead: their cell walls have undergone secondary thickening, making them impermeable to water except at the pores, called pits, in their walls. Hence, osmosis cannot be the process involved.
(d) The driving force is probably the evaporation of water from the leaves which, on account of the strong cohesion between molecules of water and their adhesion to the walls of the vessel, draws a continuous column of water through the plant. Secondly, it is possible that root pressure, resulting from an osmotic flow of water molecules across the root, may assist the ascent. A third contributory force, especially in seedlings, may be capillarity. In many herbs this force alone is sufficient to account for the ascent of sap through the xylem.
    All of these forces appear to be passive
(e) See Fig. 99.
A: xylem vessel element
B: fibre
C: medullary ray parenchyma.

21   (a) The model illustrates the mass flow hypothesis of Münch. Münch proposed that water taken from the xylem in the leaves of plants created a hydrostatic pressure that forced solutes, notably sucrose, from a 'source' in the leaves, down the phloem to a 'sink' in the roots, where osmotically active sucrose was converted into osmotically inactive starch. Münch believed that one consequence of this process was a circulation of water within the plant.
(b) (i) In order for water to circulate it would have to be drawn from xylem to phloem in the leaf and from phloem to xylem in the root. Measurements of water potential within the xylem and phloem vessels suggest that the normal movement of water is from xylem into phloem.
(ii) The use of radioactive $^{14}CO_2$ has shown that translocation of solute in the phloem is bidirectional. Moreover, it occurs at a faster rate than predictions from the mass flow hypothesis would allow.
(c) (i) Active transport
(ii) Thain has proposed that the protoplasm of sieve tubes is organised into separate strands. Some strands are believed to transport solutes in a downward direction, whilst others are believed to transport solutes upward.

22   (a) See Fig. 100.

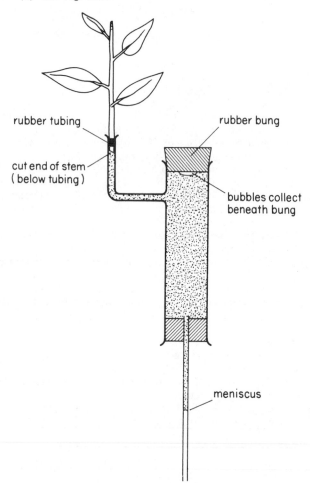

Fig. 100   Position of a cut shoot and the water meniscus in Darwin's potometer

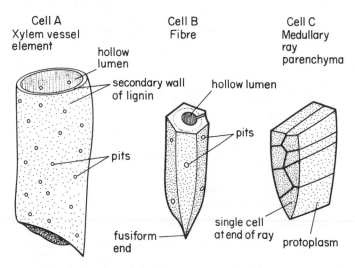

Fig. 99   Appearance of cells from xylem, as seen in a longitudinal section of tissue

(b) (i) Cut stems need to be kept in water, otherwise air bubbles may enter the xylem vessels and interfere with the ascent of water.

(ii) An airtight seal must be made around the base of the stems. If this is not achieved, the water column will fall under the influence of gravity.

(iii) Simple potometers are less accurate than some other types. It is generally advisable to allow the cut shoot to stand in position for a few minutes before attempting to take any readings.

(c) In order to demonstrate that the rate of water uptake falls as leaves are removed, rates of water uptake must be plotted against the number of leaves removed (Fig. 101). Even after all the leaves have been removed, some water is still lost through the stomata, lenticels and cuticle of the stem, as well as through the cut ends of xylem vessels.

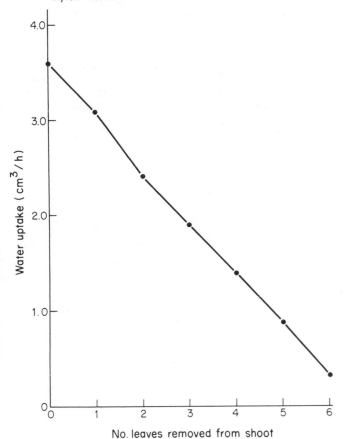

Fig. 101 The effect of removing leaves on the rate of water uptake by a cut shoot

(d) The volume of water taken up by the shoot in five minutes is found by using the formula $r^2h = 0.204$ cm³. Water uptake per hour is therefore $0.204 \times 12 = 2.45$ cm³.

In order to calculate water loss, it would be necessary to assume that water uptake and loss were identical. On that assumption, water loss per m² leaf surface is $(10\,000/263) \times 2.45 = 93.16$ cm³.

(e) Leaves were removed from the shoot and their outlines traced on graph paper mounted on thick cardboard. Leaf traces were cut from the board and weighed on a sensitive top-pan balance. Masses were compared with the mass of 10 cm², or 100 cm², mounted cardboard.

23 (a) A: water. The primary cause of stomatal opening and closing is a change in the volume of water contained in the vacuole of guard cells. As the volume of water is increased, the outer wall of each guard cell bulges outwards, causing the stoma to open.

B: light. Increased light intensity promotes the opening of stoma. This effect may occur (i) as a result of the synthesis of glucose in the chloroplasts (an event that would decrease the water potential of cell sap in the vacuoles of guard cells) or (ii) as the direct effect of light on cell membranes such as the plasmalemma, which becomes more permeable to water entering from adjacent epidermal cells.

C: chloroplasts. Illuminated chloroplasts synthesise glucose, which is an osmotically active solute, capable of reducing the water potential of guard cells to below that of adjacent epidermal cells.

D: starch gains. During daylight, and largely as a result of a change in pH, starch, which is osmotically inactive, is converted to glucose phosphate, an osmotically active solute. The reverse process is believed to occur at night, thereby increasing the water potential of the cell sap.

E: K⁺ ions. Recent theories of stomatal opening postulate that opening of the stoma is associated with the active transport of K⁺ ions, through the plasmalemma and into the guard cell.

F: plasmalemma. In addition to acting as a potassium–sodium pump, pumping K⁺ ions into the guard cells when the stomata open, the plasmalemma may be able to facilitate the active transport of water molecules.

(b) (i) Water vapour, oxygen (ii) Carbon dioxide

(c) Internal factors: an increase in the rate of photosynthesis, an increase in the rate of transpiration
External Factors: higher light intensity, a rise in temperature, an increase in wind velocity, a fall in the relative humidity of the air.

24 (a) A: erythrocytes. B: lymphocyte. C: monocyte. D: neutrophil. E: platelets.

(b) These components of blood are all constituents of the same tissue. All have arisen in bone marrow from a single type of parent cell called a megakaryocyte.

(c) Frog and bird erythrocytes possess nuclei

(d) Neutrophils, basophils and acidophils

(e) Neutrophils and monocytes

(f) Erythrocytes (g) Lymphocytes

(h) The biconcave surface increases the surface area: volume ratio, providing a large surface area for gaseous exchange.

(i) Erythrocytes (j) Erythrocytes (k) Erythrocytes

(l) Blood platelets are principally concerned with blood clotting.

Damaged blood vessels ⟶ Platelet breakdown ⟶ Thromboplastin

Prothrombin + Thromboplastin ⟶ Thrombin

Fibrinogen + Thrombin ⟶ Fibrin (forms a clot over wound)

(m) (i) Liver to kidney (ii) Muscles to kidney
(iii) Ileum to liver
(iv) Anterior pituitary to ovary or testis
(v) Liver to blood capillaries

25 (a) The pulse is a pressure change transmitted as a wave through the arterial wall and blood column to the periphery. Pulse rates are generally measured at the wrist, by placing a finger upon the radial artery.

(b) (i) The normal pulse rate in an adult human is around 70 beats per minute.

(ii) As the students anticipate the period of exercise, levels of adrenalin in their blood streams are increased. This results in a slightly increased rate of heart beat.

(iii) Student B is fitter than student A. In completing the same amount of physical work, student B's heart has to work hard to pump the required amount of blood to the tissues. Student B is also fitter in that the pulse rate has taken less time to return to the normal resting level.

(iv) There is a need to record pulse rates during the period of exercise. The shape of the curve produced would provide further information about the performance of the heart under stress.

26 (a) Blood pressure is measured with a sphygmomanometer. The instrument consists of a pressure gauge connected to an inflatable rubber cuff.

(b) The cuff is wrapped around the upper part of the arm, above the elbow. The cuff is inflated to a higher pressure than the blood pressure. A stethoscope is placed below the cuff and the operator, while deflating the cuff, listens for the sound of flowing blood.

(c) Systolic blood pressure, the minimum pressure required to completely stop blood flow in the arteries, is detected with the stethoscope at a point where, as pressure on the cuff is released, a faint tapping sound can be heard. Diastolic pressure, which is always lower than systolic pressure, is detected at the point where the tapping sounds become indistinct, or where the sound completely disappears. At this point external pressure has no constricting effect on the diameter of the arteries.

(d) (i) An increase in blood pressure, detected primarily by the pressoreceptors, is conveyed to the brain by sensory nerves. These sensory nerves synapse with motor nerves, which make connections with effectors. On receiving an appropriate stimulus, the sinus node of the heart slows the rate of heart beat, thereby reducing the pressure of blood flow from the heart. Simultaneously, circular muscles in the arteries and arterioles are relaxed, which dilates the blood vessels, reducing peripheral resistance to blood flow.

(ii) Chemoreceptors respond both to an increase in CO_2 concentration or a decrease in O_2 concentration.

(e) 1.0:3.16:16.0

(f) Cardiac output = $\dfrac{\text{Arterial blood pressure}}{\text{Peripheral resistance}}$

or Arterial blood pressure = Cardiac output × Peripheral resistance.

27 A: aorta.   B: innominate artery.   C: common carotid artery.   D: external jugular vein.   E: anterior vena cava. F: anterior mesenteric artery.   G: hepatic vein. H: iliac vein.   I: posterior vena cava.

28 Arteries: i, ii, vi, x
Veins: i, vi, vii, viii, ix, x
Capillaries: i, iii, iv, v

---

4   Modes of Nutrition and Excretion

1 (d) **2** (e) **3** (a) **4** (c) **5** (b) **6** (e)
7 (c) **8** (b) **9** (c) **10** (d) **11** (b) **12** (e)
13 (c) **14** (a) **15** (b) **16** (b) **17** (c)
18 (c) **19** (b)

20 (a) Radioactivity first appears in ribulose diphosphate (C_5) and hexose diphosphate (C_6), which suggests that the CO_2 combines with the C_5 compound to form a C_6 compound. Secondly, the formation of a C_3 compound, phosphoglyceric acid, implies splitting of the C_6 molecule. Finally, phosphoglyceraldehyde is formed by reduction of phosphoglyceric acid. At the same time, levels of ribulose diphosphate and hexose diphosphate are increased, implying a cyclic process.

(b) Oxygen is an inhibitor of photosynthesis.

(c) Atmospheric levels of CO_2 are suboptimal for photosynthesis. Rates of photosynthesis can be increased, particularly at high light intensities, by increasing concentrations of CO_2 in the atmosphere.

(d) See Fig. 102.

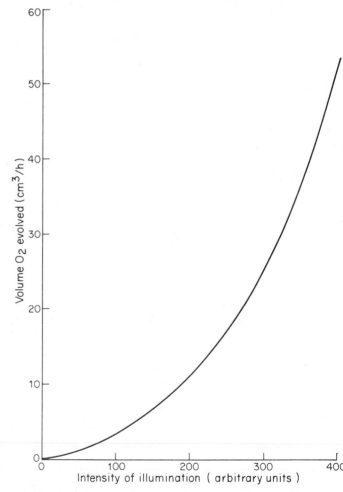

Fig. 102 The relationship between the intensity of illumination and the rate of oxygen evolution in a potted plant

(ii) An insufficient number of readings have been taken to enable the relationship between light intensity and rate of evolution of O_2 to be plotted. Additional readings, especially at high light intensities are required.
(iii) The oxygen evolved must have been derived from water.
(e) (i) Temperature (optimum 25–35°C, depending on species)
(ii) Movement of the plant, as in wind, which reduces the rate of photosynthesis

21 (a) A: ammonification. B: nitrification. C: denitrification. D: nitrogen fixation. E: nitrogen assimilation. F: leaching.
(b) X: *Nitrosomonas*. Y: *Nitrobacter*. Z: *Rhizobium*.
(c) (i) Nitrogen   (ii) NO_3^- and NH_4^+
(d) Symbiotic (mutualistic) association
(e) The red pigment leghaemoglobin may bind molecules of oxygen, thereby excluding oxygen from the anaerobic bacteria inside the nodule.
(f) Xylem
(g) (i) Root (ii) Leaf
(h) The TCA (Krebs') cycle
(i) Blue-green algae e.g. *Nostoc*

22 (a) A plant's compensation point is reached when the uptake of carbon dioxide in photosynthesis equals the rate of carbon dioxide output in respiration. If the compensation point is exceeded, a plant respires sugars faster than it can assimilate them, resulting in a loss of mass.
(b) See Fig. 103.

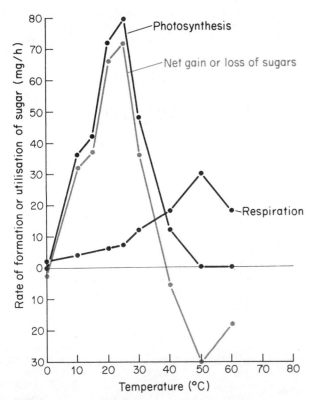

Fig. 103 The effects of temperature on photosynthesis, respiration and the rate of sugar accumulation or utilisation

(c) (i) 25°C (ii) 50°C   (d) 38°C
(e) True photosynthesis refers to the total amount of photosynthate (P) synthesised by a plant. Apparent photosynthesis refers to the actual amount of sugar accumulated by a plant: it is equal to total assimilate, minus that amount of assimilate used in respiration (R). Hence, P = true photosynthesis; $P - R$ = apparent photosynthesis.

23 (a) W: bulb   X: corm.   Y: stem tuber.   Z: tap root. A: tunic.   B: scale leaves.   C: terminal bud.   D: axillary bud.   E: telescoped stem.   F: adventitious roots.   G: node.   H: internode.   I: rhizome. J: rhizome scar.   K: cork with lenticels.   L: vascular cambium.   M: phloem.   N: xylem.   O: cortex. P: epidermis.   Q: lateral root.
(b) W: bulb. (i) Onion:fructose and water.   (ii) Tulip: starch and water.
X:corm.   (i) Crocus:starch and water.   (ii) Gladiolus: starch and water.
Y:stem tuber   (i) Potato:starch and water.   (ii) Jerusalem artichoke:inulin and water.
Z:tap root. (i) Carrot:fructose and water.   (ii) Parsnip:starch, glucose and water.
(c) Glycogen is stored in the liver. Fat is stored in depots beneath the skin, especially around the kidneys, buttocks and eyes. Protein is not stored by mammals, who require a daily dietary intake of this nutrient.
The liver stores vitamins, notably A and D, while iodine is concentrated by and stored in the thyroid gland.

24 (a) A: coronoid process.   B: articular process.   C: angle. D: carnassial tooth.   E: dentary.   F: incisor teeth. G: canine tooth.   H: premolar teeth.   I: molar teeth.
(b) See Fig. 104.

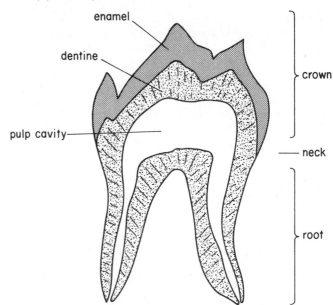

Fig. 104 Carnassial tooth of a dog (longitudinal section)

(c) In dogs the incisor teeth are small, forming a comb for cleaning the fur. Additionally, they may be used for pulling meat from bone. Canine teeth are large and pointed, used for stabbing, tearing and holding the prey. Both the premolar and molar teeth are used to cut strips of flesh from the carcase of the prey.

(d) In rabbits the incisor teeth are sharp and pointed, used for cutting grass or gnawing bark. Canines are absent: a gap in the tooth row in this position is called a diastema. Both the premolars and molars are closely packed, with a flattened surface, used for grinding and crushing food.

(e) Carnivores, in general, possess claws, have large eyes set forward in the skull, and a relatively short alimentary canal. Herbivores, on the other hand, generally possess hooves, have relatively small eyes, set laterally in the skull, and a relatively long gut, an adaptation to the fact that they must consume a large volume of plant material in order to extract an adequate amount of nutrient from it. Additionally, celloclastic (cellulose hydrolysing) bacteria may be present in one or more regions of the alimentary canal, enabling the animal to make use of dietary cellulose by converting it into glucose.

(f) $i\frac{3}{3}$, $c\frac{1}{1}$, $p\frac{4}{4}$, $m\frac{2}{3} = 42$

25 (a) A: mesentery. B: peritoneum. C: longitudinal muscle. D: circular muscle. E: submucosa. F: muscularis mucosa. G: lamina propria. H: lumen. I: epithelium.

(b) (i) Pyloric sphincter (ii) Duodenum (iii) Buccal cavity (iv) Ileum (v) Stomach (vi) Ileum (vii) Colon (viii) Duodenum (ix) Anus (x) Oesophagus

26 (a) Endopeptidases cleave specific, centrally positioned peptide bonds within polypeptides e.g. pepsin, trypsin, chymotrypsin.
Exopeptidases catalyse the removal of terminal amino acid residues from polypeptides and peptides e.g. amino peptidase, carboxypeptidase, tripeptidases and dipeptidases.

(b) Most mammalian endopeptidases are secreted in an inactive form e.g. trypsinogen, pepsinogen, which only become active in the lumen of the alimentary canal.

(c) (i) Trypsin (ii) Chymotrypsin (iii) Pancreatic juice (iv) Pancreatic juice (v) Peptidase (erepsin)

(d) Human proteins are formed from those nine essential amino acids, together with three other amino acids that the human body is able to synthesise from them.

(e) Excess amino acids are deaminated in the liver into a corresponding keto-acid and ammonia. The ammonia is combined with CO_2 in the liver to form urea, which is excreted by the kidneys.

(f) Kwashiorkor and marasmus.

27 A: gastrin: initiates secretion of HCl and pepsinogen from gastric pits
B: enterogastrone: inhibits secretion of HCl and pepsinogen from gastric pits
C: cholecystokinin: causes contraction of the gall bladder, which results in an increased flow of bile
D: secretin: stimulates the secretion of a watery, alkaline fluid from the pancreas, and bile salts from the liver and gall bladder
E: pancreozymin: stimulates the release of enzymes from the pancreas
F: duocrinin: stimulates secretion of enzymes from Brunner's glands and the Crypts of Lieberkühn
G: enterocrinin: inhibits secretion of enzymes from the jujenum and pancreas

28 Insect: ii, vii, x
Fish: i, iii, iv, v, vi, viii
Mammal: i, iii, v, ix

29 (a) A virus is a sub-microscopic, intracellular parasite that completes its life cycle within a single cell of its host.

(b) A: head. B: tail. C: capsomere. D: nucleic acid. E: core. F: tail fibres. G: tail plate.

(c) Bacteria (d) T4

(e) Protein and nucleic acid

(f) Radial, cubical and bilateral

(g) Interferon

(h) (i) Carried by insects or other vectors
(ii) Blown by wind
(iii) Washed by rain
(iv) Transferred via seeds, cuttings etc. from parent to offspring

(i) Herschey and Chase used bacteriophages with radioactive sulphur in their protein coats and radioactive phosphorus in their nucleic acids. They found that only radioactive phosphorus entered the host cell: therefore, new viral proteins must have been formed under the influence of nucleic acid.

30 (a) A: flagellum. B: mucilage. C: cell wall. D: free ribosome. E: mesosome. F: nucleic acid. G: cytoplasm. H: food reserve.

(b) (i) Cocci (ii) Bacilli (iii) Sprilla (iv) Vibrio

(c) (i) Cell walls composed of mucopeptides, proteins and lipids
(ii) DNA single-stranded, not complexed with proteins. The DNA forms a circular chromosome, not surrounded by a nuclear envelope
(iii) Ribosomes free in cytoplasm. No organelles
(iv) Transfer of genes by conjugation, transformation and transduction

(d) *Acetobacter aceti* is employed in the manufacture of vinegar and *Lactobacillus delbruckii* in the manufacture of lactic acid. *Bacillus subtilis* is a source of some enzymes added to washing powders, while *Streptococcus thermophilus* is added to pasteurised milk in the manufacture of cheddar cheese.

(e) Autotrophic bacteria are those capable of synthesising organic compounds from inorganic materials, whereas heterotrophic bacteria, including both saprophytic and parasitic forms, can feed only on organic compounds, which have been elaborated by other organisms.

31 Viruses: b, c, e, g, h, j Bacteria: a, d, f, i

32 (a) A parasite is an organism that obtains all, or some, of its foodstuffs and water from another living organism, often belonging to the same kingdom, but always to a different species. As a result of this association, the parasitised organism (host) suffers some degree of cell or tissue damage.

(b) (i) Endoparasites, such as liver flukes and tapeworms, spend their lives inside the body of their host, whereas ectoparasites, such as fleas and leeches, attach themselves to the outside of the body, feeding by puncturing the skin.
(ii) Permanent parasites, such as tapeworms, never leave the body of their host, while temporary parasites, such as fleas and mosquitoes, make intermittent visits to their host, often moving from one host to another.
(iii) An obligate parasite, such as a tapeworm, is completely dependent upon its host for all of its nutrients and water. Facultative parasites, however, such as eyebright, are capable of synthesising some of their nutrients as a result of carrying out photosynthesis in their leaves. Alternatively, like the roundworm

Rhabditis hominis, some parasites may be capable of a completely free-living existence.

(c) (i) Biological control. Attempts are made to increase the number of organisms that either feed on, or parasitise, the parasite.

(ii) Chemical control. Chemical compounds are used either to attack free-living stages in the life-cycle of the parasite, or the parasite itself in the cells or tissues of its host. In order for chemotherapy to be successful, the chemical compound must be more toxic towards the parasite than the host.

(iii) Engineering control. This involves the draining of swamps, or clearing of land, such as areas of water or water-logged soil, where stages in the life-cycle of the parasite may occur.

(d) (i) Pests spoil or damage standing crops, reducing their yield, or they attack stored grain or root crops, damaging the stored product as a result of their feeding, urination or defecation.

(ii) Predators are carnivores which devour their prey. No physiological relationships exist between predators and their prey.

(e) (i) Antibiotics (e.g. penicillin, ampicillin, tetracycline)
(ii) Mercaptan, benomyl
(iii) Benzene hexachloride, carbamyl
(iv) Metaldehyde
(v) 2, 4-D

(f) (i) Tapeworm: eyes and other sense organs are absent, along with other organs of locomotion and predation. There is no gut.

(ii) Liver fluke: each adult animal is hermaphrodite; many self-fertilised eggs are produced each day.

(iii) Liver fluke: immature forms, such as redia larvae, are capable of asexual reproduction in order to facilitate production of offspring in large numbers.

(iv) Tapeworm: hooks and suckers have evolved at the anterior end of the animal as organs of attachment. The cuticle has become modified so that the products of digestion can be absorbed through it.

(v) *Ascaris* (roundworm): capable of surviving very low oxygen tensions in gut of host, and of resisting the action of the host's digestive enzymes.

33 (a) (i) Optimal grain yields were obtained by applying 100 kg/ha nitrogen fertiliser. Yields were reduced if either smaller or larger quantities of fertiliser were applied.

(ii) Ear populations were increased by application of nitrogen fertiliser, within the range 0–150 kg/ha.

(iii) Increased application of the fertiliser had little effect on grain size, although complete absence of the fertiliser may have been size-limiting.

(iv) Optimal grain mass was achieved following application of 50–100 kg/ha fertiliser. Grain mass was reduced by applications that were either below or above this range.

(b) Grain yield = Ear population × Ear size × Grain mass
(t/ha)   (no. ears/ha)   (no grains/ear)   (g)

(c) The deeper the fertiliser is applied, the less is taken up by the plants. Hence, a surface dressing of fertiliser would be most beneficial. No attempt should be made to dig, or plough, the fertiliser into the soil.

(d) The fertiliser would probably contain ammonium (NH^+) compounds and nitrates (NO_3^-), which are the forms of nitrogen most readily assimilated by plant roots.

(e) (i) A fraction of the fertiliser, as NO_3^- and NH_4^+ ions, might remain in the soil adsorbed on the surface of soil particles.

(ii) Some of the ions, especially NO_3^-, might be leached from the soil during heavy rain.

(iii) NH_4^+ ions might be converted into NO_2^- and NO_3^- ions by the action of nitrifying bacteria e.g. *Nitrosomonas* and *Nitrobacter*.

(iv) Denitrifying bacteria might convert NO_3^- to gaseous nitrous oxide and nitrogen.

34 (a) A: spore.   B: cell wall.   C: plasmalemma.   D: cytoplasm.   E: tonoplast.   F: glycogen granule.   G: cell vacuole.   H: nuclei.

(b) Carbohydrases hydrolyse starch, glycogen and cellulose into glucose and other monosaccharides; lipases break down fats to fatty acids and glycerol; proteases break down proteins into amino acids. The end products of this extracellular digestion dissolve in a film of water that surrounds the hyphae, are absorbed into the hyphae, and elaborated into the body of the fungus.

(c) Mould fungi, such as *Mucor*, are decomposers, which break down the remains of dead plants and animals, that might otherwise litter the surface of the earth.

(d) Some fungi, such as late blight of potatoes, are parasitic, causing heavy losses of crops. Others form mycorrhizal associations with the roots of higher plants, an association which probably enables the plant to obtain more mineral salts from the soil.

(e) (i) Mushrooms are grown for use as a vegetable.
(ii) Yeasts are used in brewing and baking.
(iii) *Penicillium chrysogenum* is the source of an antibiotic.
(iv) *Penicillium roquefortii* is employed in the manufacture of Roquefort cheese.

5   Respiration and Respiratory Systems

1 (e)   2 (c)   3 (c)   4 (b)   5 (c)   6 (b)
7 (b)   8 (c)   9 (a)   10 (b)   11 (d)   12 (a)
13 (b)   14 (b)   15 (e)   16 (b)   17 (d)
18 (d)   19 (c)   20 (d)

21 (a) (i) Anaerobic respiration, alternatively called glycolysis, is the conversion of a molecule of glucose by a sequence of enzyme-catalysed reactions to two molecules of pyruvic acid. Reduction of the pyruvic acid results in the formation of ethanol in plants and lactic acid in animals.

(ii) Aerobic respiration is the oxidation of pyruvic acid, produced in glycolysis, to CO_2 and H_2O, involving enzyme-catalysed reactions of the Krebs' cycle and the transfer of electrons, from reduced hydrogen acceptors, along an electron transport chain.

(b) (i) Cytoplasm   (ii) Mitochondria

(c) A: glycolysis.   B: alcoholic fermentation.   C: lactic acid fermentation.   D: Krebs' cycle.   E: oxidative phosphorylation.

(d) (i) Green plants, yeasts
(ii) Animals, certain bacteria

(e) Compounds in pathway E act as hydrogen acceptors.

(f) Citric, α-ketoglutaric, succinic, malic

(g) (i) 2   (ii) 36

22 (a)
$$RQ = \frac{\text{Moles of } CO_2 \text{ produced}}{\text{Moles of } O_2 \text{ absorbed}}$$

(b) (i) 1.0   (ii) ∞   (iii) 0.69   (iv) 1.33   (v) 4.0

(c) Warburg apparatus, Dixon–Barcroft simple respirometer

(d) (i) Alkaline pyragallol
(ii) Concentrated (5 M) NaOH or KOH

(e) RQ determinations give a fairly reliable index of the nature of the substrate being respired, providing all of the respiration is aerobic and only carbohydrates, fats and proteins are being utilised (RQ glucose = 1.0, fat = 0.7, protein = 0.9). But if organic acids are being respired, or if some of the respiration is anaerobic, figures in excess of 1.0 do not provide a reliable index of the nature of the material being respired.

(f) (i) 0.737 litres   (ii) 1.412 litres   (iii) 0.829 litres

(g) (i) 17.44 kJ   (ii) 39.38 kJ   (iii) 17.4 kJ

(h) 179.09 litres

23 (a) Sigmoid

(b) During exercise, Following a rise in levels of CO_2

(c) (i) 28% at pH 7.5, and 36% at pH 7.2
(ii) Displacement of the oxygen dissociation curve to the right is called the Bohr shift. The effect is important because, at any given oxygen tension, more oxygen is released from the pigment and made available to the tissues. A rise in body temperature could produce a similar effect.

(d) (i) 17.3 kPa   (ii) 100%

(e) (i) Pulmonary vein   (ii) Pulmonary artery

(f) Dissociation curves (i), (ii) and (iii) would be drawn to the left of curve X. The higher oxygen affinity of foetal haemoglobin helps in the transfer of oxygen to the foetal blood in the placenta. Similarly, the higher oxygen affinity of myoglobin helps in the transfer of oxygen at the muscles. As the mud-dwelling animal lives under conditions of low oxygen tension, its haemoglobin also has a high affinity for oxygen.

(g) Haemoglobin in earthworms and certain insects e.g. *Chironomus* Haemocyanin in snails and cuttlefish

24 (a) A: expiratory centre.   B: inspiratory centre.
C: respiratory centre.   D: sensory nerve.
E: motor nerves.

(b) The rate of breathing in a mammal is controlled by the respiratory centre in the medulla oblongata of the brain. This region is composed of two parts, an inspiratory centre and an expiratory centre, only one of which can function at a particular time. Motor nerves from the expiratory centre innervate both the diaphragm and intercostal muscles. When an impulse passes along these motor nerves it causes contraction of the intercostal muscles and diaphragm, which initiate inspiration. Stretch receptors in the lungs are connected, via sensory nerves, to the expiratory centre. Stimulation of the stretch receptors at the end of inspiration therefore triggers relaxation of muscles of the diaphragm and costal region, leading to expiration.

(c) (i) Rise in blood CO_2, rise in pH, deficiency of O_2, rise in temperature
(ii) Rise in blood pressure

(d) See Fig. 105. During exercise an oxygen debt is incurred, which is repaid after exercise has ceased. The magnitude of this debt and the repayment, indicated by shaded areas on the graph, are identical.

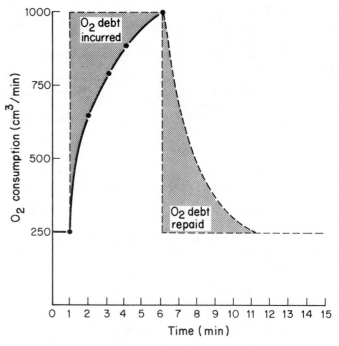

Fig. 105   Oxygen consumption during modest exercise

(e) Rise in CO_2

(f) (i) Accumulation of lactic acid in the muscles
(ii) Depletion of ATP reserves

25 (a) (i) Stoma in leaves: diffusion
Lenticels in trunk and branches: diffusion.
(ii) Tracheae: diffusion
(iii) Skin: transport in blood, combined with haemoglobin in the plasma
(iv) Gills: transport in the blood, combined with haemoglobin in red corpuscles
(v) Skin: diffusion
Lungs: transport in blood, combined with haemoglobin in red corpuscles

(b) Oxygen is obtained from the surrounding water as a result of the following sequence of events.
(i) The mouth and oral valve are opened, followed by a lowering of the floor of the mouth cavity. This creates a sucking force that draws water into the mouth.
(ii) The mouth, oral valve and opening to the oesophagus are closed, then the floor of the mouth cavity is raised, forcing a jet of water outwards via the gill slits, positioned between the gills and covered by the operculum.

(c) A respiratory membrane must be thin, moist and permeable to gases.

(d) In small animals, such as *Amoeba*, the ratio of surface area to volume is large. Oxygen can diffuse through the thin cell membrane to reach all parts of the cell. Similarly, carbon dioxide can diffuse from the animal into the surrounding water. There is therefore no need for a respiratory system.

(e) In green plants oxygen reaches all of the tissues by diffusion, whereas in vertebrates there is a muscular pump, associated either with the gills or lungs, which pumps oxygen to the respiring cells. Moreover, in vertebrates, oxygen combines with a carrier, haemoglobin, at the respiratory surface and is released from this carrier in the tissues. Plants have no vascular system for the transportation of oxygen to their tissues.

26 (a) 73.5% N_2, 13.8% O_2, 4.6% CO_2, 8.1% H_2O
(b) Most of the CO_2 and H_2O entering the alveolus are excretory products, derived from the oxidation of carbohydrates, fats and proteins in the tissues.
(c) CO_2 may reach the lungs in three different forms, or combinations:
(i) The greatest proportion of CO_2 in the blood is in the form of bicarbonate—sodium bicarbonate in the plasma and potassium bicarbonate in the red cells.
(ii) In combination with proteins, as carbamino-proteins. Haemoglobin, for example, may carry some CO_2 in this form.
(iii) As carbonic acid in solution—although the acid generally dissociates into bicarbonate ions and hydrogen ions.
(d) Red blood corpuscles contain the enzyme carbonic anhydrase, which accelerates the reaction:

$$CO_2 + H_2O \rightleftharpoons H_2CO_3$$

If this enzyme were not present, the reaction could not occur during the short time the blood is in the capillaries.
(e) (i) Folds in the walls of the alveolus provide a large surface area for gaseous exchange. Walls of healthy alveoli are elastic.
(ii) Any breakdown in the structure of the alveolus would reduce the surface area for gaseous exchange, forcing the heart to pump ever larger quantities of blood to the lungs. This would place an added strain on the heart and might lead to heart failure.
(f) Victims of drowning are given oxygen for the needs of their tissues, and CO_2 to stimulate the respiratory centre of the brain, in an attempt to induce inspiration.

27 (a) Spirometer
(b) See Fig. 106.
(c) (i) Breathe out as deeply as possible
(ii) Breathe in as deeply as possible
(iii) Resume normal breathing
(d) (i) 5.7 litres   (ii) 1.5 litres   (iii) 1.5 litres
(iv) 0.4 litres

(e) (i) The rate of breathing (ventilation) would be increased
(ii) The tidal volume of the lungs would be increased

28 (a) A: spiracular valve.   B: bristle hairs.   C: trachea.
D: tracheoles.   E: fluid in base of tracheoles.
F: muscle cells.
(b) Structure A is a two-lipped valve, which can open or close to allow gaseous exchange between the insect and the atmosphere. Additionally, the valves regulate water loss by evaporation from the body.
Structure B is a bristle hair. These occur around the spiracle, and serve to filter out dust and foreign bodies from the air before it enters the tracheal system.
(c) (i) The spiracular valves would be open when the insect is active. In many insects opening occurs in response to an increase in levels of CO_2.
(ii) The spiracular valves would close during periods of rest or hibernation, or when the insect was placed in a particularly arid atmosphere.
(d) Air diffuses through the spiracles, passes along the tracheae and into the intracellular tracheoles. At their extreme ends, tracheoles are filled with liquid. Oxygen dissolves in this fluid and then diffuses into nearby cells.
(e) Chitin, in spiralling strands, lines the tracheae. Unlike the cuticle of the insect, the lining of the tracheal system is not shed during ecdysis.
(f) (i) Tracheae must remain open if they are to be efficient in bringing oxygen to the tissues. Rigidity of the tracheal walls is imparted by spiralling strands of chitin. If the size of insects were to be scaled up, these strands of chitin would bend and become deformed, buckling under increased pressures imposed by larger mass.
(ii) Many insects have no means of pumping gases in or out of their tracheal system. Where diffusion is the only means of supplying oxygen to the tissues, any increase in the size of the tracheal system, would lead to ever decreasing efficiency in the supply of oxygen.

29 (a) (i) 0.5M: 23g ethanol   (ii) 1.0M: 46g ethanol
(iii) 1.5M: 69g ethanol   (iv) 2.0M: 92g ethanol
(v) 2.5M: 115g ethanol   (vi) 3.0M: 138g ethanol
(b) (i) 0.5M: 5.83% ethanol   (ii) 1.0M: 11.6% ethanol
(iii) 1.5M: 17.5% ethanol   (iv) 2.0M: 23.3% ethanol
(v) 2.5M: 291% ethanol   (vi) 3.0M 350% ethanol
(c) See Fig. 107.

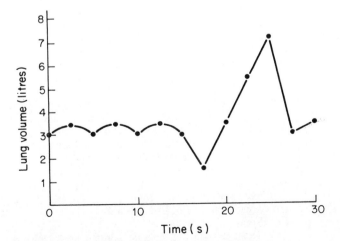

Fig. 106 Kymograph trace of respiratory movements in a normal human subject

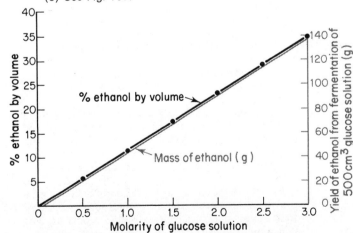

Fig. 107 Theoretical yields, by mass and volume, of ethanol from the fermentation of solutions of glucose

(d) (i) 113g glucose   (ii) 177g glucose
(e) (i) A hydrometer
    (ii) Actual yields of ethanol would be lower than predicted yields, as the yeasts use some of the sugar as a source of C, H and O for their own structural carbohydrates, fats and proteins, from which the protoplasm of daughter cells is composed.
(f) During fermentation of sugars by yeast, considerable volumes of CO_2 are evolved. If fermentation vessels are sealed, the pressure of the evolving gas may cause glass vessels to burst, scattering glass over a wide area. The danger can be avoided by using specially constructed thick-walled fermentation vessels, fitted with air locks.

6 Reproduction, Development and Growth

1	(c)	**2**	(c)	**3**	(d)	**4**	(c)	**5**	(c)	**6**	(a)
7	(d)	**8**	(e)	**9**	(a)	**10**	(d)	**11**	(b)	**12**	(c)
13	(b)	**14**	(e)	**15**	(b)	**16**	(e)	**17**	(e)		
18	(c)	**19**	(c)	**20**	(a)	**21**	(a)	**22**	(a).		

23 (a) Dioecious   (b) Monoecious   (c) Protandrous   (d) Protogynous   (e) Micropyle   (f) Male gametes   (g) Ovum   (h) Embryo   (i) Primary endosperm nucleus   (j) Endosperm   (k) Basal cell   (l) Terminal   (m) Plumule   (n) Radicle

24 (a) Flower A: $K_5 C_5 A_5 G_{\overline{1}}$
       Flower B: $P_{3+3} A_3 G_{\overline{3}}$
(b) See Fig. 108.
(c) (i) Free central   (ii) Axial
(d) (i) Introrse   (ii) Extrorse   (iii) Pendulous
(e) (i) Dicotyledonae   (ii) Monocotyledonae

25 (a) (i) Pollination is the transfer of pollen from the stamen of a flower to the stigma, either of the same flower or of a different flower.
    (ii) Fertilisation is the fusion of male and female gametes to form a zygote.
(b) Pollination may be effected by   (i) wind   (ii) insects   (iii) water, or   (iv) a self-pollinating mechanism.
(c) Anemophilous means wind pollinated and entomophilous means insect pollinated.
(d) Pollen from anemophilous flowers is small, smooth, light and often possesses air sacs, which act as sails. Pollen from entomophilous flowers is larger, sticky, and often covered by fine hairs, which adhere to the body of an insect.
(e) Pollen grains of *Amaryllis* and primrose will germinate in water only. Pollen grains of a number of species will germinate if placed into a solution of sucrose. The majority of plant species, however, require additional nutrients and probably also growth substances, secreted by the style, before their pollen grains will germinate. In these species the direction of pollen tube growth is influenced by chemotropic substances. In a fourth group of plants, typified by cherry, pollen grains will germinate only on the stigmas of flowers which are genetically compatible.
(f) Mitosis

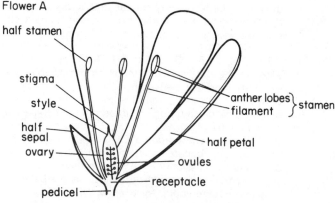

Flower A

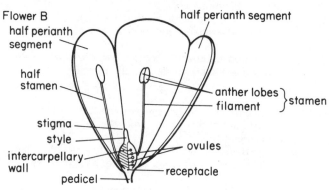

Flower B

Fig. 108   Appearance of flowers A and B in vertical section

26 (a) A: style   B: ovary wall   C: micropyle   D: integuments   E: funicle   U: antipodal cells   V: polar nuclei   W: ovum   X: synergidae   Y: tube cell   Z: male gametes.
(b) A → B → C
(c) B: ovary (fruit) wall   D: testa of seed   E: hilum of seed.   V: endosperm.   W: embryo.

27 (a) Hazel, plantain   (b) Primrose, loosestrife   (c) Hazel, cucumber   (d) Berry, drupe   (e) Broad bean, sunflower   (f) Maize, castor oil   (g) Sunflower, mustard   (h) Broad bean, pea   (i) Frog, sticleback   (j) Oestrogen, progesterone   (k) Prostate, Cowper's glands   (l) Aphids, honeybee   (m) Butterfly, housefly   (n) Amnion, chorion   (o) Ectoderm, mesoderm   (p) Periblem, pleurome   (q) Gibberellin, auxin   (r) Temperature, light intensity   (s) Somatomedin, testosterone.

28 (a) (i) 28 days   (ii) Day 14   (iii) Fallopian tube   (iv) Uterus   (v) Identical twins form from the same egg, which divides into two separate cells following its first cleavage. Unidentical twins form from separate eggs, which are fertilised simultaneously.
(b) A: umbilical cord   B: amnion   C: chorion   D: amniotic fluid   E: umbilical artery and vein   F: uterus   G: placenta.
(c) (i) Water, amino acids and glucose
    (ii) Proteins, hormones
    (iii) The placenta has a large surface area: volume ratio. Material and embryonic bloods are brought into close contact, separated only by a thin membrane

across which diffusion can take place. Some of the membranes which separate maternal from embryonic blood may be capable of active transport of materials, notably nutrients.

(iv) The gut, kidneys and lungs

(v) Progesterone, which directly inhibits FSH secretion from the anterior pituitary gland, indirectly prevents development of new ovarian follicles. Additionally, progesterone maintains the endometrium in a condition necessary for the successful retention of an implanted embryo.

Other hormonal secretions of the placenta include large quantities of oestrogen, together with human chorionic gonadotrophin (HCG), a luteinising hormone, and human placental lactogen which stimulates development of the mammary glands in readiness for milk production.

(vi) Marsupials

(d) (i) Progesterone

(ii) One effect of progesterone is fluid retention, which may increase a woman's mass by several kilograms. This additional mass may impose a strain on the heart, particularly in women who are already obese. Secondly, progesterone may accelerate the rate of blood clotting, increasing the possibility of thromboembolism or other disorders of the blood-clotting mechanism.

(iii) Progesterone, oestrogen, chorionic gonadotrophin and placental lactogen all stimulate the development of the mammary glands during pregnancy.

(iv) Teats, such as occur in sheep and cattle, contain many ducts which secrete milk into a central reservoir, which is drained by one opening to the surface.

Nipples, such as occur in humans, have many small ducts, each with its own small terminal reservoir, opening on to the surface of the nipple.

(v) Prolactin, secreted by the anterior pituitary, initiates the milk flow.

(vi) Milk is a liquid food for the young, which contains a balanced diet of carbohydrates, fats, proteins, mineral salts, vitamins and water.

(vii) Rennin coagulates milk, thereby slowing its passage through the alimentary canal of a young mammal.

$$\text{Milk} \xrightarrow{\text{Rennin}} \text{Curd} + \text{Whey protein}$$

29 (a) Oogenesis   (b) Zygote   (c) Cleavage
(d) Animal pole   (e) Blastocoel   (f) Gastrulation
(g) Archenteron   (h) Notochord   (i) Neurulation
(j) Ectoderm   (k) Amniotic fluid   (l) Organisers
(m) Allantois   (n) Foetus   (o) Induction
(p) Dorsal lip of blastopore

30 (a) (i) Metamorphosis is a change in form, from juvenile stage to adult, that occurs in the life cycle of certain animals.

(ii) Frog

(b) (i) Hemimetabolous metamorphosis, or incomplete metamorphosis, occurs in insects such as the locust, in which the juvenile stage resembles the adult. There are three stages in the life cycle:

Egg → Nymph → Imago

(ii) Holometabolous metamorphosis, or complete metamorphosis, occurs in insects, such as butterflies,

in which juvenile stages bear no resemblance to the adult. There are four stages in the life cycle:

Egg → Larva → Pupa → Imago

(c) The larva is slow moving and therefore more likely than the adult to be attacked by predators.

(ii) The larva has a different food source from the adult. This allows the insect to utilise two or more different food sources during its life history.

(d) See Fig. 109.

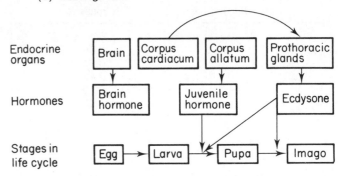

Fig. 109 Endocrine organs, hormones and stages in the life cycle of a butterfly

(e) (i) A: hatching of eggs. B: larval moult. C: pupal moult.

(ii) Ecdysis

(iii) The animal is holometabolous as it has two major ecdyses in its life cycle, these relating to larval–pupal and pupal–imago metamorphoses.

(iv) Juvenile hormone (neotenin)

(v) See Fig. 110.

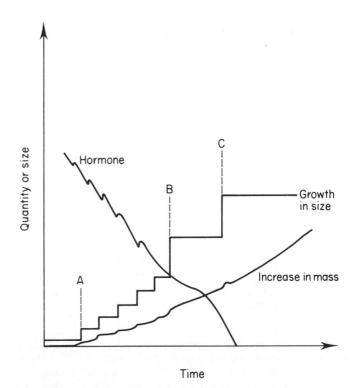

Fig. 110 Graph illustrating growth in size of an insect and the production of growth hormone

31 (a) A: stigma   B: aleurone   layer   C: cotyledon
        D: coleoptile   E: plumule   F: hilum
        G: coleorhiza   H: radicle   I: scutellum
        J: endosperm   K: fused pericarp and testa.
     (b) (i) Table 49 shows the total dry mass of the whole
         grain.
         (ii) See Fig. 111.

Table 49

Time (days)	Dry mass (g)
0	0.47
2	0.46
4	0.44
6	0.42
8	0.40
10	0.42

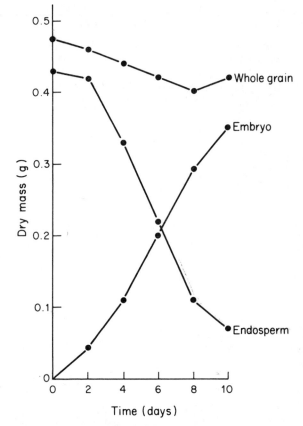

Fig. 111 Changes in the dry mass of a whole maize grain,
endosperm and embryo during germination

     (c) Before germination of a cereal grain can occur, water
         absorbed into the seed by *imbibition* must carry
         gibberellic acid from the *embryo* to the *aleurone layer*,
         where it stimulates *enzyme induction* of *α-amylase*.
         This enzyme, released into the *endosperm* hydrolyses
         starch into *maltose*. Further hydrolysis of maltose
         produces mobile products, such as *glucose and fruc-
         tose*, which pass across the *scutellum* into the
         embryo, where they promote growth of the *plumule*
         and *radicle*.

     (d) (i) Two, or four, plants were removed from a field of
         maize during each week of the growing season. The
         plants were dried and weighed. Mean masses were
         plotted as a graph.
         (ii) The graph is sigmoid.
         (iii) The grand period of growth.
         (iv) Initial growth of the seedling is at the expense of
         stored food, which is respired, causing an initial loss in
         mass. Once the seedling has broken the surface of the
         soil, however, photosynthesis occurs, and there is from
         that point (point A) an increase in dry mass, resulting
         from an accumulation of the products of
         photosynthesis.
         (v) The final loss in mass results from removal of
         seeds, which either fall from the plant or are eaten by
         birds and other animals.

32 (a) (i) C   (ii) B   (iii) A   (iv) C   (v) C
     (b) (i) *Mucor*   (ii) Moss   (iii) Frog
     (c) Fig. 112.
     (d) A diploid organism possesses a pair of each of its
         genes. As a result, diploid organisms do not necessarily
         express all of the recessive genes that they carry,
         whereas haploid organisms express all of their recess-
         ive genes.

(i) Detailed life cycle of a fern (type B)

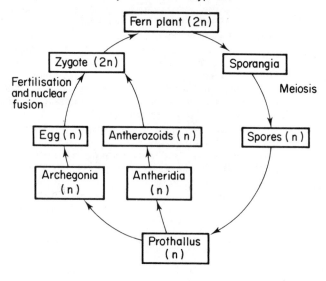

(ii) Detailed life cycle of a human (type C)

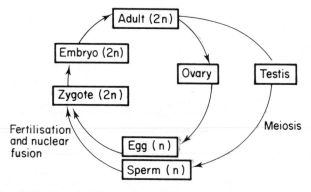

Fig. 112 Detailed life cycles of fern and man

7 Support, Movement and Dispersal

1 (e) **2** (a) **3** (b) **4** (a) **5** (c) **6** (e)
7 (b) **8** (c) **9** (a) **10** (e) **11** (b) **12** (c)
13 (a) **14** (a)

15 (a) A: epidermis. B: cortex. C: medullary ray. D: pith. E: phloem. F: vascular cambium. G: epidermis. H: exodermis. I: cortex. J: endodermis.

(b) W: angular collenchyma. X: pericyclic fibres. Y: metaxylem fibres. Z: metaxylem fibres.

(c) (i) A peripheral cylinder of mechanical tissue is formed from angular collenchyma. Mechanical tissue in the vascular bundles is organised into longitudinal strands, with strong, relatively inflexible fibres positioned on each side of soft, relatively flexible phloem, forming an I-girder.

(ii) Mechanical tissue in the root is in the form of a narrow central strand. The mass of stems and leaves produces a downward compression on the stem, which is supported against the force of gravity by the peripheral cylinder of mechanical tissue. Additionally, the vascular strands, which resemble vertical columns, bear some of the mass of this downward compression.

Lateral movements of the stem, which occur in wind, are made possible by the provision of I-girders in the vascular bundles, yet the bundles have sufficient tensile strength to resist breakage.

In roots the central strand of mechanical tissue resists the force of extension.

(d) (i) The vascular bundle is extended between bundles to form a cylinder.

(ii) Concentric rings of xylem and phloem are formed by the vascular bundle.

(iii) The epidermis is replaced by bark.

(iv) Xylem vessels at the centre of the stem may become blocked by deposits of resin (tyloses), forming 'heartwood'.

(e) Secondary thickening increases the amount of mechanical tissue in the stem, enabling the stem to support the considerable mass of branches and leaves. The formation of 'heartwood' creates, in effect, a hard central post, resistant to downward compression. A thick, rigid, fibrous bark replaces a thin, soft, flexible epidermis.

16 (a) A: articular cartilage. B: spongy bone. C: epiphysis. D: calcified cartilage. E: diaphysis. F: periosteum. G: compact bone. H: marrow.

(b) (i) Testosterone and oestrogen, produced during adolescence, stimulate calcification of cartilage between the epiphysis and diaphysis, preventing further growth in the length of the bone.

(ii) The kidneys secrete erythropoietin, which stimulates red blood cell formation in bone marrow.

(iii) Calcitonin from the thyroid gland increases bone calcium accretion.

(iv) Parathyroid hormone mobilises calcium from the bones.

(v) Vitamin D, either formed in the skin or ingested in the diet, regulates the supply of calcium to bone, chiefly by affecting calcium absorption across the intestinal mucosa.

(vi) Vitamin B_{12}, cyanocobalamin, is essential for red blood cell formation in marrow.

(c) (i) J: lamella of matrix. K: Sharpley's fibre. L: canaliculi. M: Haversian canal. N: Volkmann canal. O: interstitial bone.

(ii) In region G (compact bone)

(iii) L: osteocytes.
M: veins, arteries, lymph vessels.

17 (a) A: caudal vertebrae. B: sacrum. C: lumbar vertebrae. D: thoracic vertebrae. E: axis. F: atlas. G: orbit. H: nares. I: zygnemic arch. J: scapula. K: humerus. L: sternum. M: femur. N: pelvic girdle (half).

(b) (i) Axial skeleton: skull and vertebral column

(ii) Appendicular skeleton: bones of limbs, pectoral girdle and pelvic girdle

(iii) Cartilage bones are preceded in the embryo by a cartilage model, which is later replaced by bone.

(iv) Membrane bones are not pre-modelled in cartilage, but develop directly from mesenchyme.

(c) See Fig. 113.

(d) A rabbit has only four digits in its hind foot.

Fore limb   Hind limb

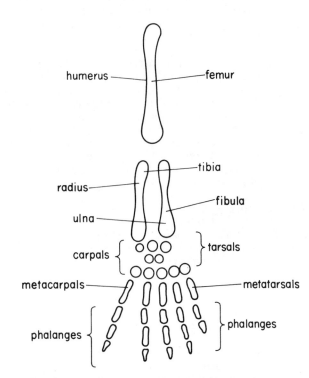

Fig. 113 Generalised structure of a pentadactyl limb

18 (a) A: neural spine. B: transverse process. C: neural arch. D: pre-zygopophysis. E: neural canal. F: bony centrum. G: dorsal root. H: ventral root. I: spinal nerve. J: intervertebral disc.

(b) The intervertebral discs permit movement between consecutive vertebrae, and provide flexibility to the spinal column as a whole. Additionally, the discs surround and protect the spinal nerves, cushioning them from mechanical damage that might otherwise result from movement of the bones.

(c) Muscles attached between successive neural spines bend the spine backwards (or upwards) when they contract.

Muscles attached between successive transverse processes contract to effect lateral movements of the spine.

(d) Vertebrae

(i) Vertebrae surround and protect the spinal cord.

(ii) Vertebrae serve as an attachment for muscles.

(iii) Vertebrae form a flexible girder, which, in part, supports the mass of the body against the force of gravity.

19 (a) A: marrow.   B: fibrous capsule.   C: synovial membrane.   D: synovial fluid.   E: articular cartilage.   F: tendon.   G: ligamant.   H: periosteum.

(b) Synarthrodial joints. They unite bones of the skull.

(c) (i) Between skull and atlas vertebra

(ii) In wrist and ankle

(iii) Between axis and atlas vertebrae

(iv) In shoulder and hip

(v) In elbow and knee

(d) (i) 1st order: splenius capitis muscle, attached to back of head, contracting to lift head.

(ii) 2nd order: gastrocnemius muscle, attached to ankle bone, contracting to raise body on ball of foot.

(iii) 3rd order: biceps muscle, attached to radius bone, contracting to raise forearm.

(e) Length = 30 cm, mass = 28.8 kg

Bone, the material from which the skeleton is composed, has a particular tensile strength that does not increase as the dimensions of the animal are scaled up. Therefore, as mass increases, there is a directly related increase in the compression force acting on bones. This could lead to a fracture of certain bones, or lead to excessive wear on the joints. The effect could, to some extent, be counteracted by the development of short, thick limb bones, which would, however, slow the rate of locomotion.

20 Bone: b, c, e, g, j

Cartilage: a, d, f, h, i

(k) (i) Trachea (ii) Articular cartilage

(iii) Epiglottis (iv) Pinna of ear

(v) Fibrocartilage

21 (a) A: seta or hair.   B: epicuticle.   C: exocuticle.   D: endocuticle.   E: epidermis.   F: basement membrane.   G: striated muscle.   H: tendon.   I: pore.

(b) The arthropod skeleton differs from that of a mammal in that it is

(i) composed of chitin (c.f. bone),

(ii) an exoskeleton (c.f. endoskeleton),

(iii) incapable of growth in thickness or length.

(c) (i) The epicuticle contains waxes that are very effective in preventing water loss, by evaporation, from the body.

(ii) The cuticle is tough and fairly resistant to mechanical injury.

(iii) The cuticle may be coloured, either as camouflage concealing the insect against its background, or it may be brightly coloured, warning off predators.

(d) As the endoskeleton cannot grow, it must periodically be shed in order to allow for growth. Immediately after shedding the old cuticle, the new cuticle remains soft for several hours until the processes of hardening and tanning are complete. When the cuticle is soft, and the animal immobile, it may fall an easy prey to predators.

(e) Ecdysis. The hormone ecdysone causes the endocuticle to liquify.

(f) J: coxa.   K: trochanter.   L: femur.   M: tibia.   N: tarsus.

(g) X is the antennal comb, used for removing pollen from the antenna. Y is the eye brush.

(h) Tracheae

22 (a) A: sarcomere.   B: myosin.   C: actin.   D: Z-disc.   E: A-band.   F: H-zone.   G: I-band.

(b) Each sarcomere, in turn, shortens as actin molecules slide inwards. Myosin molecules retain their position. It is believed that cross bridges form between actin and myosin molecules, and that movement is achieved as a result of a cyclic back and forth movement of the cross-bridges, making and breaking contact with sites on the actin filaments and sliding them inwards.

(c) The relative lengths of the I-bands and A-bands

(d) In isotonic contraction a muscle, while exerting a pull, is allowed to shorten.

In isometric contraction a pulling force is exerted as, for example, against a spring, but the muscle is not allowed to shorten.

(e) Use of the term 'syncytium' implies that striated muscle fibres are filamentous and multinucleate, without subdivision into separate component cells.

(f) (i) Cardiac muscle is composed of separate branched cells. Each cell is uninucleate, with cross striations.

(ii) Smooth muscle is composed of separate fusiform cells. Each cell is uninucleate, without cross striations.

23 (a) See Fig. 114.

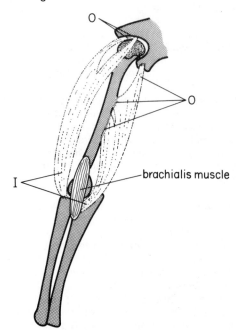

Fig. 114   Position of the brachialis muscle

(b) The origin of a muscle is the relatively fixed end, at which least movement occurs. The insertion of a muscle is its relatively movable end, the point at which the pull is greatest (see also Fig. 114).

(c) (i) The arm will straighten from a bent position.
(ii) The part of the leg below the knee will be raised.
(iii) The sole of the foot will be raised from the ground.

(d) The biceps and triceps muscles exert their pull on opposite faces of bone across the elbow joint. Contraction of the biceps bends the arm about the elbow joint, whilst contraction of the triceps straightens it. These two muscles therefore exert opposite, or antagonistic, effects.

(e) When the biceps muscle is contracted, an inhibitory neurone in the spinal cord inhibits the motor neuron that causes the triceps muscle to contract. Similarly, contraction of the triceps muscle inhibits the biceps muscle from simultaneous contraction.

24 (a) A: lateral line system.   B: pectoral fin.   C: pelvic girdle.   D: ventral fin.   E: caudal fin.   F: dorsal fin. G: swim bladder.

(b) (i) A, the lateral line system, involved in the detection and localisation of objects in the water, enables the fish to avoid collisions. Additionally, it may be sensitive to the rates of movement, both of the fish, and of the surrounding water currents.
(ii) G, the swim bladder, in which the volume of gas can be varied, adjusts the density of the fish to that of the surrounding water. This enables the fish to maintain a particular level in the water.

(c) Propulsive thrust is generated by side to side movements of the caudal fin. Dorsal and ventral fins act as stabilisers, preventing rolling. Pectoral and pelvic fins, held horizontally, generate lift, or they may be held more or less vertically to act as brakes, reducing the speed of locomotion.

(d) The body is streamlined in shape; that is, bluntly rounded anteriorly and tapering posteriorly, with maximum girth at one-third of the distance between nose and tail. When an animal of this shape moves through water, the fluid flows smoothly over the surface of its body, parting in front of it and closing behind it, thereby reducing drag.

(e) Elasmobranch fish, which have no swim bladder, are denser than the water in which they live. Once they have stopped swimming, they sink to the floor of the sea bed. An asymmetrical tail fin, with a large lobe, generates lift when swimming is resumed.

25 (a) and (b) See Fig. 115.
(c) Contraction of the vertical flight muscles flattens the upper, or dorsal, surface of the thorax, causing the wings to be raised. Similarly, contraction of the horizontal muscles restores the thorax to its original position, causing the wings to be lowered.

(d) A: lift.   B: drag.   C: gravity.   D: forward thrust.
(e) (i) Wings are present.
(ii) The insect becomes streamlined when in the air, with the legs held close to the body surface.
(f) (i) Mass has been reduced by the development of hollow bones.
(ii) The wings·have been extended to form air sacs, which occupy spaces between internal organs.

26 (a) Rain water   (b) Tsetse fly   (c) Anopheline mosquito   (d) Insects   (e) Small mammals (on fur) (f) Fleas and mosquitoes   (g) Dogs, foxes, badgers   (h) Gulf stream currents   (i) Flies (j) Wind   (k) Aphids and sap-sucking bugs

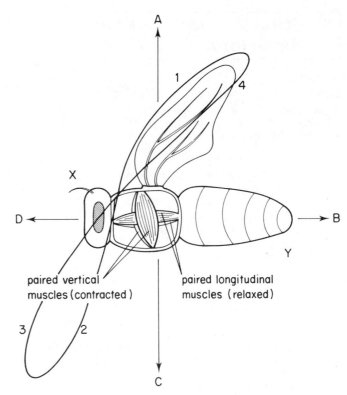

Fig. 115   Diagrammatic representation of an insect in flight

(l) Bacteriophages   (m) Explosive self-dispersal (n) Wind   (o) Wood-boring beetles

27 (a) A: remains of style and stigma.   B: placenta. C: funicle.   D: ovule (or seed).   E: calyx. F: receptacle.   G: pedicel.

(b) Hairs on the seed pod may:
(i) help to reduce water loss from the seed pod,
(ii) help to protect the young, green pod from damage by high light intensity.

(c) From the limited number of observations made, it would appear that seeds of gorse are dispersed during the mornings and afternoons of hot days, when the air temperature has reached 20°C, or more; but not on cool days, or at night, when the temperature is below 20°C. In addition, it may be that (i) relative humidity of the air, and (ii) light intensity, play some role in determining the time of seed dispersal. Further information is required before it is possible to reach any firm conclusions.

(d) The tabulated results could be presented visually as histograms:
(i) No. pods exploding (y-axis) against time of day (x-axis)
(ii) No. pods exploding (y-axis) against mean air temperatures (x-axis).
Results constitute a frequency distribution diagram, in which the number of pods exploding is determined within a certain range of (i) time and (ii) air temperature.

(e) Ripe, dry pods explode more rapidly than moist, unripe pods. Therefore, in the gorse, the most immature seed pods are carried near the apex, while the older, more mature pods are carried further down the stem.

(f) Broom; geranium

8 Stimuli, Responses and Behaviour

1 (a) **2** (d) **3** (d) **4** (c) **5** (e) **6** (c)
7 (d) **8** (c) **9** (c) **10** (c) **11** (e) **12** (e)
13 (b) **14** (e) **15** (b) **16** (e) **17** (a)
18 (b) **19** (e) **20** (e) **21** (d).

22 (a) (i) See Fig. 116.
(ii) The plumule is positively phototrophic and the radicle negatively phototropic. Even so, the response to unilateral light is more marked in the plumule than in the radicle, which shows a relatively small curvature compared with that of the plumule.

(b) A bioassay is a measurement of the concentration of a particular compound, such as auxin, by determining its effects on living tissue.

(c) In order to bioassay amounts of auxin synthesised in the excised coleoptiles of maize, the excised tips are allowed to stand on blocks of agar for a standard period of time. Blocks of uniform size are then transferred to young seedlings of barley, from which the coleoptile tips have been removed, leaving the plumule in a central position. With the block resting on one side of the erect shoot, subsequent curvature during a standard period of time is measured.

(d) (i) Maximum auxin production occurs in darkness. The effect of uniform illumination is to reduce auxin production by 0.9%
(ii) When the illuminated and dark sides of the coleoptile tip are completely separated from one another by a glass cover-slip, both halves synthesise almost equal amounts of auxin, with auxin production on the illuminated side slightly in excess of that on the dark side. When separation of the halves of the coleoptile tip is incomplete, however, more auxin is collected on the dark side, implying lateral translocation of auxin, from the illuminated to the dark side, under the influence of unilateral illumination.
(iii) In darkness, 29% of the auxin is translocated laterally. Lateral translocation is increased away from unilateral illumination, but slightly inhibited towards it.

(e) (i) These results imply that a majority of the seedlings, but not all, made a positive hydrotropic response, bending toward the moist area of agar, in response to a moisture gradient in air.
(ii) The ability of seedlings to respond to a moisture gradient appears to be more marked in some species than in others.

(f) Bending of the radicles in the direction of the wooden rod may have been caused by any of the following factors, acting singly or in combination.
(i) The wood may have been wetter than the glass.
(ii) The surface of the wood and the surface of the glass may have had different pH values.
(iii) The wood may have presented a rough surface, whereas the glass surface was smooth.
(iv) The wood may have absorbed mineral ions, or chemical compounds, that influenced the growth rate of the radicle.

(g) Positive hydrotropism in radicles does not appear to be a universal response, made by all seedlings of all species.

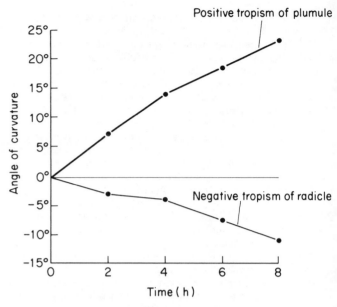

Fig. 116 Curvature of a plumule and radicle of a maize seedling in response to unilateral illumination

(i) Positive geotropism is a universal response, made by the radicles of all seedlings.
(ii) Negative phototropism of the radicle is shown in some, but not in all, species. In those species where it does occur, however, all members of the species show the same response.

23 (a) (i) Inhibited: stimulated. (ii) Stimulated: inhibited. (iii) Inhibited: stimulated. (iv) Inhibited: stimulated. (v) Stimulated: inhibited. (vi) Stimulated: inhibited.

(b) Phytochrome, which occurs chiefly in leaves.

(c) Phytochrome exists in two interconvertible forms; one absorbs red light at an absorption peak of 660 nm, while the other absorbs far-red light with a peak at 730 nm. When red light is absorbed by P660 it is rapidly converted into P730. P730 therefore tends to accumulate in plants placed in sunlight. Normally, P730 slowly reverts to P660 during the hours of darkness, but a rapid reconversion can be effected by a single flash of far-red light.

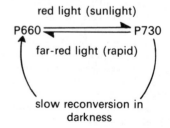

(d) It seems unlikely that phytochrome has a direct influence on flowering. Most likely, it initiates production of a hormone, or florigen, that is translocated from leaves to the apical regions of stems, where it initiates the formation of flower bud primordia.

(e) Different definitions have been given but the most widely accepted definitions are as follows.
(i) Short-day plants produce flowers in response to short periods of illumination, less than 10 hours in each

24-hour period. e.g. chrysanthemum, Michaelmas daisy.

(ii) Long-day plants produce flowers in response to long periods of illumination, exceeding 14 hours in each 24-hour period. e.g. wheat, lettuce.

(iii) Day-neutral plants produce flowers regardless of the length of photoperiod to which they are exposed. e.g. cucumber, tomato.

Table 50

Daylength (h)		Flowering response	
Light	Dark	Species A	Species B
8	16	✓	
9	17	✓	
16	8		✓
17	9		✓

24 (a) Homeostasis is the maintenance of a constant internal environment in the face of changes in the external environment.

(b) (i) Endotherms (homiotherms)
(ii) Ectotherms (poikilotherms)

(c) (i) Surface capillaries dilate; hair erector muscles are relaxed; sweat glands secrete sweat.
(ii) Surface capillaries constrict; hair erector muscles contract, raising the hair follicles to a more erect position; sweat glands stop secreting sweat.

(d) Adrenalin, noradrenalin, thyroxin

(e) (i) Shivering
(ii) Running, or any form of activity that would increase the rate of heat production in skeletal muscles

(f) (i) Babies have a large surface area relative to their volume. Additionally, they are unable to generate heat by shivering.
(ii) In old people the skin capillaries often lose their ability to constrict. As these capillaries may be permanently dilated, more heat tends to be lost through the skin surface than in adolescent or middle-aged members of the population.

(g) (i) Insulin from the β-cells of the pancreatic islets of Langerhans reduces blood glucose, whereas glucagon from the α-cells of the same organ increases levels of glucose in the blood.
(ii) Adrenalin, secreted by the adrenal medulla, increases levels of glucose in the blood.
(iii) Glucocorticoids, a product of the adrenal cortex, raise levels of glucose in the blood.
(iv) Adrenocorticotrophic hormone (ACTH), from the anterior pituitary gland, stimulates secretion of aldosterone from the adrenal cortex.
(v) Aldosterone increases salt uptake into the bloodstream.
(vi) Antidiuretic hormone (ADH), secreted by the posterior pituitary gland, increases water absorption into the blood from the kidney; urine volume decreases and urine concentration increases.

25 (a) A: dendrites.   B: cell body.   C: myelin sheath.   D: node of Ranvier.   E: axon.   F: synapse.   G: mitochondria.   H: vesicles.   I: post synaptic membrane, with receptor sites.

(b) Autonomic nervous system

(c) Motor neurons occur in the ventral roots of the peripheral nervous system.

(d) A sensory neuron would not bear a terminal cell body, and it would conduct an impulse from a sense organ to a part of the central nervous system. Additionally, a sensory neuron would generally have longer dendrites and a shorter axon than a motor neuron.

(e) All excitable cells, including neurons, have a threshold membrane potential for the initiation of an action potential. To initiate an action potential, the stimulus must depolarise the membrane to its threshold level. Once depolarised, the wave of depolarisation passes along the entire length of the neuron. There is therefore a full transmission of an impulse along an axon, or no transmission.

(f) When the motor neuron terminal is invaded by the action potential, synaptic vesicles release acetylcholine, or another transmitter, into the synaptic cleft. The neurotransmitter attaches to receptors embedded in the postsynaptic membrane, triggering an action potential in the next neuron in the series. As soon as the action potential has been triggered, an enzyme, cholinesterase, breaks down the neurotransmitter into choline and acetate.

(g) (i) A neurotransmitter is a chemical compound, secreted at a synapse, that either promotes or inhibits transmission of an impulse to the next neuron in the series.
(ii) Acetylcholine, dopamine.

(h) J: sensory neuron.   K: relay neuron.   L: motor neuron.   M: dorsal root ganglion.   N: white matter.   O: grey matter.   P: spinal canal.

(i) Reflex arc, on the left hand side of the body

(j) Reflex action

(k) Knee-jerk reflex; emission of semen in a male

(l) Region N is composed of white matter, whereas region O is composed of grey matter. Whilst both regions are composed of similar neurons, in white matter one is viewing chiefly axons which run longitudinally and have been cut transversely. In grey matter, however, one is viewing both axons and cell bodies that run transversely and have been cut longitudinally.

26 (a) A: cornea.   B: aqueous humor.   C: iris diaphragm.   D: ciliary muscle.   E: lens ligaments.   F: sclerotic.   G: choroid.   H: retina.   I: vitreous humor.   J: blind spot.   K: optic nerve.

(b) See Fig. 117.

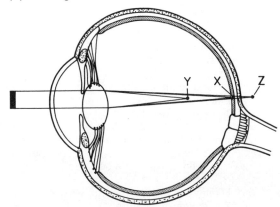

Fig. 117 Paths of light rays through the eye in faultless vision (X), myopia (Y) and hypermetropia (Z)

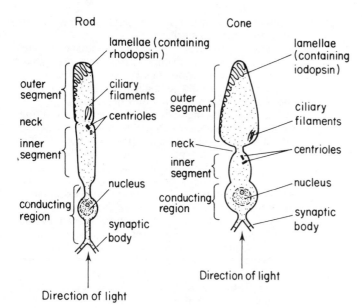

Fig. 118   A rod and cone

(c) See Fig. 118.
(d) (i) Rods   (ii) Cones   (iii) Rods   (iv) Cones
    (v) Rods
(e) Group L: (i) Basilar membrane of cochlea, (ii) sound,
    (iii) in response to a sound wave of particular
    frequency, the basilar membrane moves upward to
    make contact with the overlying tectorial membrane.
    Group M: (i) Gelatinous cupule in ampulla of semi-
    circular canal, (ii) rotational movements, (iii) the
    cupula is displaced by the flow of fluid in the semi-
    circular canals.
    Group N: (i) Maculae of utricle and saccule,
    (ii) gravity, (iii) movements of the calcareous otoconia
    under the influence of gravity stimulate the cilia of hair
    cells.
    Group O: (i) Maculae of utricle and saccule, (ii)
    acceleration, (iii) acceleration causes deflection of the
    stereocilia towards the kinocilium.

27 (a) A hormone is a soluble chemical compound of low
       molecular mass, produced by an endocrine gland and
       transported by the blood stream to a target organ,
       where it exerts its effects.
   (b) (i) Insulin, growth hormone
       (ii) Testosterone, oestrogen
   (c) (i) Insulin and glucagon
       (ii) Gibberellin and abscisic acid
   (d) (i) Anterior pituitary: testis and ovary
       (ii) Posterior pituitary: uterus during labour
       (iii) Anterior pituitary: thyroid
       (iv) Kidney: bone marrow
       (v) Duodenum: pancreas
   (e) Vasopressin, ADH and oxytocin are all formed in nerve
       cells near the hypothalamus and are transported along
       axons to the nerve endings in the neuropophysis.
   (f) (i) Prevention of leaf, petal and fruit drop
       (ii) Promotion of root formation
       (iii) Elimination of broad-leaved weeds
       (iv) Prevention of sprouting
       (v) Production of seedless grapes
       (vi) Increased α-amylase secretion

28 (a) Stretch reflex; emission of semen
   (b) 'Eye' spots on wings of butterfly; red underside of male
       stickleback
   (c) Bees; wasps
   (d) Bees; migrating birds
   (e) Rods; cones
   (f) Halteres of housefly; cochlea of ear
   (g) Photonasty; thermonasty
   (h) Geotropism; phototropism
   (i) Oat; turnip
   (j) Chrysanthemum; michaelmas daisy

29 Male: d, g, i, f, j, a
   Female: h, e, b, c

30 (a) Sign stimuli   (b) Circardian rhythm
   (c) Habituation   (d) Instrumental conditioning
   (e) Classical reflex conditioning   (f) Imprinting
   (g) Insight learning, or reasoning   (h) Displacement
   activity   (i) Instinct   (j) Kinesis   (k) Orientation by
   stellar-compass   (l) Reflex

9   Ecology

1 (b)   2 (b)   3 (e)   4 (a)   5 (a)   6 (c)
7 (c)   8 (b)   9 (a)   10 (b)   11 (c)   12 (c)
13 (d)   14 (c)   15 (b)   16 (b)   17 (e)
18 (c)   19 (c)   20 (c).

21 (a) (i) $GP = NP + R_1$ (ii) $NP = GP - R_1$
       (iii) $SP_1 = A_1 - R_2$ (iv) $A_1 = SP_1 + R_2$
       (v) $NA_1 = I_1 - A_1$ (vi) $A_2 = SP_2 + R_3$
   (b) Fig. 66 shows that of (i) the total incident radiation
       falling on a leaf,
       (ii) 17% is reflected, (iii) 13% is absorbed and
       (iv) 70% transmitted.
   (c) (i) 1.15%   (ii) 8.6%   (iii) 7.5%
       (iv) 4313kJ $/m^2/$ day   (v) 21.4%
       (vi) Decomposers (vii) Herbivores

22 (a) Climatic: a, d, l, m
       Edaphic: b, f, k
       Biotic: c, e, g, h, i, j, n

23 (a) (i) Pioneer   (ii) Climax   (iii) Succession
   (b) The age of plants could be determined by cutting stems
       at ground level and by counting the number of rings of
       xylem.
   (c) (i) Waterlogging of the soil, or trampling
       (ii) Animal grazing, or trampling
       (iii) Fire
   (d) (i) B   (ii) C   (iii) D   (iv) A   (v) A   (vi) D
       (vii) D   (viii) D   (ix) A
   (e) P: bracken (non-flowering leaves break surface in April
       and die back in October)
       Q: pine (evergreen, flowers May–June)
       R: birch (flowers before leaves appear; leaves shed in
       October)
       S: heather (evergreen, flowers August–September)
       T: gorse (evergreen, flowers February–November)

24 (a) Population A, which first appears in March, shows
       fluctuating growth, reaches a peak in August, and
       declines to extinction in October. Population B first

appears in April, shows fluctuating growth, rises to a peak in late August, and declines to extinction in late October.

(b) Graphs A and B illustrate typical relationships between numbers of a prey population (A) and its predator population (B). The density of the population of prey always exceeds that of the predator, which appears later and disappears later than the prey. Moreover, the two populations oscillate with one another, out of phase. Any increase in the population density of the prey is followed by an increase in the population density of the predator. Similarly, any decrease in the population density of the prey is followed by a corresponding fall in the population density of the predator.

(c) The number of insects in population A is most likely to be determined by times of hatching from pupae, air temperature, humidity of the air, abundance of food, immigration and emigration. The density of population B is determined largely by the numbers of individuals in population A.

(d) Level K_1 is the carrying capacity of the environment for population A; level K_2 is the carrying capacity of the environment for population B.
(i) If these levels are exceeded, populations will tend to fall back to the carrying capacity.
(ii) If population densities fall below this level, they subsequently rise, to the level of the carrying capacity.

(e) Population growth curves A and B are seasonal, while the growth curve of the human population is not seasonal. The peak population density of populations A and B are determined by climatic factors, which have little, if any, effect on human population. The human population is not subject to control by predation.

25 (a) See Fig. 119.

(b) The size of the bacterial population is limited chiefly by availability of the food supply. In addition, some waste products of metabolism, such as alcohols, or acids, may exert a toxic effect and slow the rate of growth of the population.

(c) As population density of *Tribolium confusum* increases, so adult beetles consume a larger percentage of eggs, thereby providing a control to further increase in the size of the population.

(d) See Fig. 120. In order to make the figures comparable, it is necessary to calculate the number of female beetles per 800 grains as shown in Table 51.

In the grain weevil *Sitophilus oryzae* the number of eggs laid per female decreases as the density of the population increases. This may occur
(i) as females may hesitate to lay in grains already containing eggs or larvae, or
(ii) because insects may hinder one another as density increases, and so prevent egg laying.

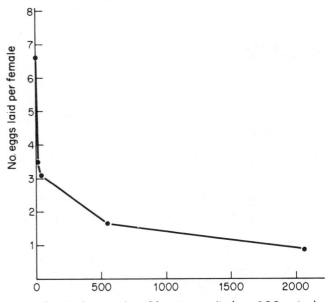

Fig. 120 The relationship between the population density of female grain weevils and the number of eggs laid per female

Table 51

No. female weevils per 800 grams	Mean no. eggs laid per female
4	6.6
16	3.5
64	3.02
512	1.6
2048	0.9

26 (a) (i) Transect   (ii) Quadrat

(b) The transect would be used to mark equidistant points, spaced one or two metres apart, across the shore. The quadrat, with 50 or 100 squares, would be used to determine the density of sea weeds and periwinkles at different levels. With the central point of the quadrat

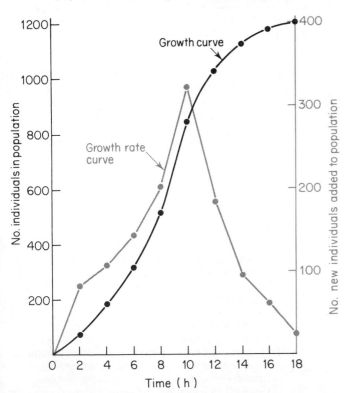

Fig. 119 The population growth curve and growth rate curve of a bacterium

positioned beneath a mark on the transect, densities would be determined by:

(i) percentage of squares containing fronds of a particular sea weed,

(ii) no. of periwinkles occurring within the area enclosed by the quadrat.

(c) Before using the apparatus, it would be necessary to identify the species of sea weeds and the species of periwinkle present on the shore. This information could be obtained by the use of keys, and checked by reference to texts containing coloured photographs or drawings of the organisms.

(d) See Fig 121. The distribution of *Littorina neritoides* parallels that of *Enteromorpha intestinalis*; and that of *Littorina littorea* parallels the distribution of *Fucus serratus*. *Littorina rudis* and *Littorina obtusata* occur in a zone where the distribution of *Ulva lactuca* and *Fucus serratus* overlap.

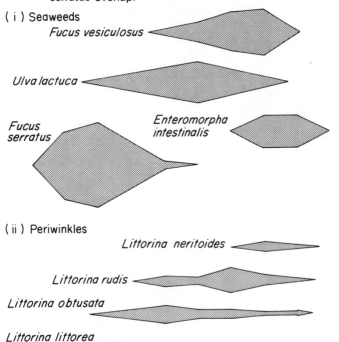

Fig. 121 Histogram showing the distribution of seaweeds and periwinkles on a seashore

(e) Food preferences of the edible periwinkle could be investigated:

(i) by dissecting specimens collected from the shore and analysing the contents of their alimentary canals, or

(ii) by offering measured masses of different sea weeds to periwinkles, and recording the mass of each plant consumed.

27 (a) See Fig. 122.

(b) (i) See Table 52. The optimal density of sowing is 16 plants per m².

(ii) At densities below 16 plants per m² the land is under-utilised. At densities above 16 plants per m²

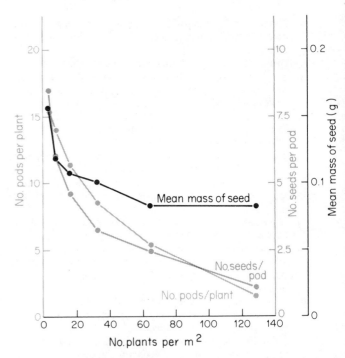

Fig. 122 The relationship between the number of plants per square metre, the number of pods per plant, the number of seeds per pod and the mean mass of the seeds

Table 52

No. plants/m²	Total mass seed harvested from each plot(g)
4	70.21
8	86.89
16	104.75
32	85.3
64	77.5
128	22.1

plants were probably prevented from reaching their full size, possibly on account of insufficient nutrients being absorbed by the roots, or insufficient light reaching the leaves. Additionally, as the population density of the plants increases, the relative humidity between plants increases, and these conditions favour the growth of parasitic fungi, which attack the plants and reduce the yield.

(c) Bean plants and weeds would have been competing for soil water and mineral salts, as well as for light.

(ii) Bean plants may have been successful in this competition largely on account of their size, which would give them an advantage in competition with smaller plants, and possibly on account of having deeper, more spreading root systems, which would obtain mineral salts and water from a larger area of the soil. If nitrate concentrations in the soil were limiting, the presence of root nodules in the bean, containing bacteria that are capable of fixing nitrogen from the atmosphere, would have been of some advantage.

(d) The growth of wheat is unaffected by the presence of poppy. On the other hand, poppy is much more successful when sown with wheat than when sown alone. This is probably related to the type of environment provided by the wheat as it grows, an environment that is shaded and moist.

(e) The growth of barley is not affected by the presence of chickweed. On the other hand, the presence of barley greatly reduces the productivity of chickweed and the number of flowers per plant. This effect is believed to occur because the barley produces a chemical compound from its roots which inhibits the growth of chickweed.

28 (a) 550–600

(b) See Fig. 123. The approximate size of the population is estimated at 575.

(c) (i) 80%   (ii) 20%

(iii) 'Loss rate' would be a better term that 'death rate'. The 20% reduction in the population may have been caused by emigration, not necessarily by death.

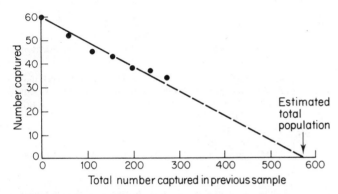

Fig. 123 An estimate of total population size from removal sampling

29 (a) From A (i) fine mud that would cover the leaves of plants, reducing the efficiency of their photosynthesis, and interfere with the respiration of gill-breathing animals, such as fish,

and (ii) fertilisers that would promote the growth of algae, which, in turn, would slow the flow rate of the river.

From B, (i) fertilisers would act as above, and (ii) herbicides that would have a toxic effect on rooted plants growing at the river's edge, and possibly also exert a toxic effect on animal life.

From C (i) detergents that are toxic to both plant and animal life, and

(ii) hot water, which may kill both plants and animals immediately below the outfall.

From D sewage, which reduces the oxygen content of the water, but adds nutrients that promote the growth of bacteria, fungi and algae.

From E metallic objects, which rust in the water. Rust, apart from its toxicity to many plants and animals, also reduces the oxygen content of the water.

From F oil, lead and rubber from the road surface, all of which are toxic to plants and animals.

(b) A measurement of the biochemical oxygen demand (BOD), generally defined as the mass of dissolved oxygen (mg) required by a definite volume of liquid (1 litre) during five days' incubation at 20°C, would determine if the river was polluted, particularly by the growth of microorganisms.

(c) Raw sewage contains coliform bacteria, especially *E. coli*. Water samples are mixed either with a nutrient fluid medium containing bile salts, lactose and a pH indicator, or applied to the surface of an agar gel, containing the same constituents. Following incubation, the presence of *E. coli* may be detected by a change in the colour of the pH indicator, from blue-purple to yellow, as lactic acid is formed.

(d) (i) (G) mineralisation, (H) reduction, (I) assimilation, (J) putrefaction, (K) oxidation

(ii) Putrefaction and reduction of SO_4^{2-} to H_2S are most likely to occur in a polluted river.

(iii) Green plants and bacteria

(iv) Cysteine and methionine

30 (a) See Fig. 124.

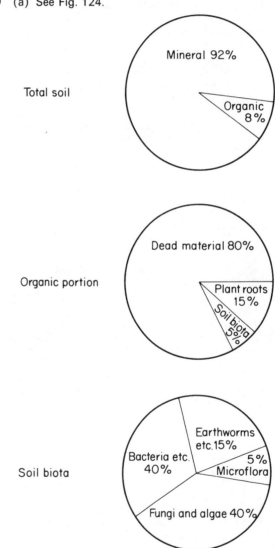

Fig. 124 Percentages of living and non-living components of garden soil

(b) (i) 0.02–2.0 mm : 0.002–0.02 mm

(ii) high : low

(iii) relatively dry and warm : relatively cold and wet

(iv) dry and dusty : hard and cracked

(v) low : high

(vi) causes erosion : little, if any, effect

(c) Earthworms in the soil: (i) bury leaf litter and raise mineral particles, derived from the weathering of rocks, to the surface, (ii) shred dead leaves as a result of feeding on leaf litter, (iii) mix humus and mineral particles into aggregates, (iv) improve the aeration of the soil as a result of burrowing, and (v) neutralise the acidity of material passed through their digestive tracts.

(d) (i) Ash from wheat stubble contains large amounts of calcium carbonate, which is alkaline.

(ii) Rain leaches mineral salts, including carbonates, from the surface of the soil. As alkaline carbonates are washed out of the soil, the pH becomes more acid, often returning to its original level within 2–3 weeks, depending on rainfall.

(iii) The advantages of burning wheat stubble include the destruction of pests, including weed seeds, immediate return of minerals, as soluble salts, to the soil, and the neutralisation of soil acidity.

(iv) Apart from the risk of fire spreading to hedges etc., stubble-burning may be destructive of earthworms and useful insects. Again, it may deprive the soil of humus, which is an important agent in maintaining the crumb-structure of the soil.

(e) (i) Nitrogen fixers (ii) Denitrifiers (iii) *Nitrobacter* (iv) *Nitrosomonas* (v) Ammonifiers (vi) Ammonia urea, uric acid and trimethylamine oxide

10 Genetics

1	(b)	2	(c)	3	(e)	4	(a)	5	(b)	6	(e)
7	(b)	8	(b)	9	(d)	10	(b)	11	(d)	12	(a)
13	(c)	14	(d)	15	(a)	16	(b)	17	(e)
18	(b)	19	(a)	20	(e).

21 (a) (i) Promotor (ii) Operator (iii) Structural (iv) Regulator (v) Repressor (vi) Inducers (vii) Inducible

(b) Maltose induces the production of maltase.

(c) The dark brown pigment, present in the tips of the ears, tail etc, develops only in the cooler extremities of the body. Higher temperatures prevent expression of this gene.

22 (a) Supergene: a cluster of closely linked genes, which are inherited as a single unit.

(b) Mutation: a spontaneous change in a chromosome or gene.

(c) Telocentric chromosome: a chromosome in which the centromere is at or very near one end, so that the arms are of unequal length.

(d) Karyotype: the chromosomal complement of an individual organism.

(e) Epistasis: the condition in which the effect of one gene, or gene pair, is masked by the action of another gene.

(f) Apomixis: the omission of meiosis and fertilisation from a sexual cycle: sexual reproduction is often replaced by asexual reproduction.

(g) Pleiotrophy: a single gene exerts its effects on a number of different characters.

(h) Position effect: the position of a gene on a chromosome may vary and the effect of the gene may differ, according to its position.

(i) Aneuploidy: variation in the chromosome number, generally by $+1$ or -1.

(j) Bivalent: two homologous chromosomes joined together as in the zygotene and pachytene stages of prophase and metaphase I of meiosis.

(k) Euploidy: variations in the chromosome number, consisting entirely of multiples of a single basic number.

(l) Parthenogenesis: development of an egg without fertilisation.

(m) Non-disjunction: a failure of homologous chromosomes to segregate during metaphase I of meiosis.

(n) Probability: a prediction, based on chances between 0% and 100% that a particular event will take place.

23 (a) A_2 and A_3

(b) (i) Antipodal cells (ii) Synergidae (iii) Tube (vegetative) nucleus

(c) A_1a (d) A_2A_3a (e) (i) A_1A (ii) A_2A_3A

(f) (i) Self-fertilisation helps to ensure that eggs are not wasted, as each fertilised egg generally develops into an F_1 individual.

(ii) One of the disadvantages of self-fertilisation is that individuals arising after several generations of selfing are homozygous. In fact, heterozygosity is reduced by one-half at each generation.

(g) (i) 50% (ii) 75% (iii) 87.5%

(h) (i) All male flowers (ii) 1 male: 1 female flower

24 (a) A: male nuclei. B: tube nucleus. C: antipodal nuclei. D: polar nuclei. E: ovum. F: synergidae.

(b) AB, Ab, aB, ab

(c) AB

(d) (i) A or a (ii) B or b (iii) AB, Ab, aB, ab (iv) B or b (v) B or b (vi) A or a (viii) AB, Ab, aB, ab (viii) AA or aa (ix) AAB, AAb, aaB, aab

(e) ABC, ABc, AbC, Abc aBC, aBc, abC, abc

25 (a) (i) Self pollination is prevented by the arrangement of stamens and carpels, which are widely separated and at different levels in the flower, so that transfer of pollen from stamen to stigma is unlikely to occur, and by incompatibility of the pollen, as an ss grain will not germinate on a ss stigma, nor will SS pollen germinate on a SS stigma.

(ii)

Pollen		Ovule	Offspring
Ss	$\times$	ss	$1Ss:1ss$
ss	$\times$	Ss	$1Ss:1ss$

(b) (i) S_1S_2 (ii) S_2S_3, S_2S_4 (iii) $S_1S_3, S_1S_4, S_2S_3, S_2S_4$ (iv) S_2S_3, S_2S_4, S_3S_4 (v) None

(c) (i) 50% (ii) 50% (iii) 100% (iv) 75% (v) None

26 (a) Linked genes are located on the same chromosome.

(b) As linked genes occur on the same chromosome, they tend to stay together, and not undergo independent assortment during the formation of gametes.

(c) A backcross $AaBb \times aabb$, would be made to test for linkage. If the genes are on separate chromosomes the F_1 genotypes will be $\frac{1}{4} Aabb: \frac{1}{4} Aabb: \frac{1}{4} aaBb$ and $\frac{1}{4} aabb$, but if the two genes are linked there will be significantly more parental types ($AaBb$ and $aabb$) than predicted, and significantly fewer recombinants ($Aabb$ and $aaBb$).

(d) See Fig. 125.

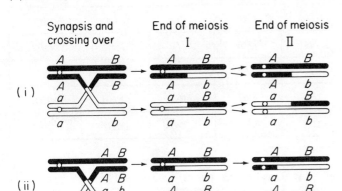

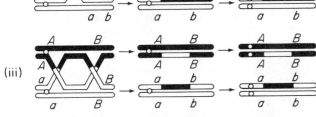

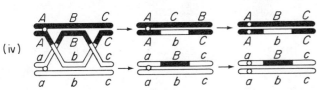

Fig. 125 Synapsis and crossing over within a pair of homologous chromosomes

(e) Crossover Value = $\dfrac{\text{No. recombinant offspring} \times 100}{\text{Total number offspring}}$

(f)

$A \qquad E \qquad\qquad D \qquad C \qquad B$

$\underline{4844}$

(g) $AB = 20$ $AD = 12$ $DB = 8$ $AE = 4$ $EC = 12$

27 (i) P cross = $Aa \times aa$ F_1 cross = $aa \times aa$
The gene for blackness is completely dominant to the gene for whiteness; there is no sex linkage.
(ii) P cross = $Aa \times Aa$
F_1 cross = $Aa \times aa$ or $Aa \times AA$
F_2 cross = $Aa \times Aa$
Gene A is incompletely dominant to gene a; there is no sex linkage.
(iii) P cross = $Aa \times Aa$
F_1 cross = $Aa \times aa$ or $Aa \times AA$
F_2 cross = $Aa \times aa$ or $Aa \times AA$
F_3 cross = $Aa \times aa$ or $Aa \times AA$
The black character is expressed only in homozygous dominant or recessive form; it is not sex-linked.
(iv) P cross = Aa or aa or $Aa \times aa$
$\quad$ F_1 cross = $Aa \times Aa$
$\quad$ F_2 cross = $Aa \times aa$ or $Aa \times AA$
The gene is sex-linked and expressed in males in homozygous form, either aa or AA.

(v) P cross = Aa or Aa
$\quad$ F_1 cross = $AA \times Aa$ or $aa \times Aa$
$\quad$ F_2 cross = $AA \times Aa$ or $aa \times Aa$
The gene is sex-linked and expressed in females in homozygous or heterozygous form, either AA, Aa or aa, Aa.

28 (a) See Fig. 126.

Human
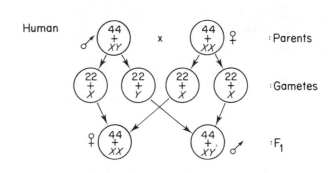

Chicken
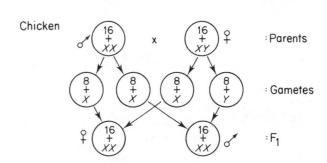

Grasshopper
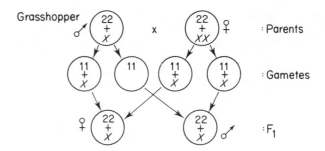

Honeybee
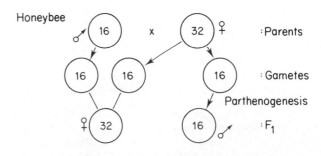

Fig. 126 Sex inheritance in the F_1 generation of different organisms

(b) See Fig. 127.
(c) (i) Non-disjunction in anaphase I of meiosis
    (ii) Aneuploid
(d) (i) Kleinfelter's syndrome
    (ii) Turner's syndrome
    (iii) Down's syndrome

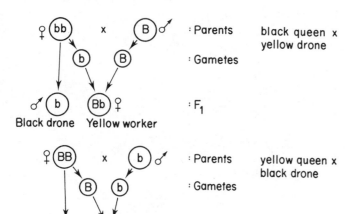

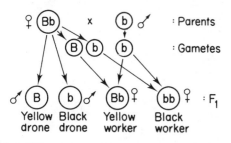

Fig. 127   Colour inheritance in honeybees

29 (a) (i) *TT* × *TT*, *TT* × *Tt* or *TT* × *tt*
       (ii) *Tt* × *tt*   *(iii)* *Tt* × *Tt*   *(iv)* *tt* × *tt*
   (b) 9 tall red: 3 tall white: 3 dwarf red: 1 dwarf white (see Table 53).
   (c) (i) 6 intermediate red: 3 tall red: 3 dwarf red: 2 intermediate white: 1 tall white: 1 dwarf white.
       (ii) 6 tall pink: 3 tall red: 3 tall white: 2 dwarf pink: 1 short red: 1 short white.
       (iii) 4 intermediate pink: 2 tall pink: 2 intermediate red: 2 intermediate white: 2 dwarf pink: 1 tall red: 1 tall white: 1 intermediate white: 1 dwarf red: 1 dwarf white.
   (d) (i) 9/16 individuals possess gene *T* in dominant form combined with gene *R* in dominant form; 7/16 individuals do not possess both genes *T* and *R* in dominant form.
       (ii) 1/16 individuals, either *TTRR* or *ttrr*, possesses both genes in homozygous dominant or homozygous recessive form. The 15/16 other individuals do not possess both genes in homozygous form.
       (iii) 12/16 individuals possess either gene *T* or gene *R* in heterozygous from (e.g. *Tt* or *Rr*); 4/16 individuals are not heterozygous for either character.
       4/16 individuals possess gene *T* or *R* in homozygous dominant or recessive form; 12/16 individuals do not possess gene *T* or *R* in homozygous dominant or recessive form.

Table 53

	TR	Tr	tR	tr
TR	TTRR	TTRr	TtRR	TtRr
Tr	TTRr	TTrr	TtRr	Ttrr
tR	TtRR	TtRr	ttRR	ttRr
tr	TtRr	Ttrr	ttRr	ttrr

30 (a) 9 oval tubers, resistant
       3 oval tubers, susceptible
       3 round tubers, resistant
       1 round tubers, susceptible
   (b) (i) 25%   (ii) 25%   (iii) 100%

31 (a) Parental genotypes: *bbSs* × *BbSs*
   (b) Parental genotypes: *bbss* × *BbSs*

32 (a) (i) *AA* × *BB*   *AA* × *Ba*   *Aa* × *BB*   *Aa* × *Ba*
       (ii) *AA* × *aa*   *Aa* × *aa*
       (iii) *BB* × *aa*   *Ba* × *aa*
       (iv) *BB* × *BB*   *BB* × *Ba*   *Ba* × *Ba*
       (v) *AB* × *aa*
   (b) *AA* × *BB*   *AA* × *aa*   *BB* × *aa*   *BB* × *Ba*
       BB × *BB*
   (c) *AA* × *Ba*   *Aa* × *BB*   *Aa* × *aa*   *Ba* × *aa*
   (d) None

33 (a) *rrrr* = white
       rrrR = light red
       rrRR = medium red
       rRRR = medium-dark red
       RRRR = dark red
   (b) *RRRR*, *rrrr*, R_1R_1rr, R_2R_2rr
   (c) (i) *RRRR* × *rrrr*   (ii) *Rrrr* × *rrrr*
       (iii) *RrRr* × *rrrr*   (iv) *rrRr* × *RRrr*

34 (a) (i) Mother either *aA* or *aa*. The genotype of the father would not have affected the inheritance in a male child.
       (ii) 50%   (iii) 50%
   (b) (i) 50%
       (ii) For a woman to show symptoms of haemophilia she would have to be homozygous for the recessive gene. This means that her father would have to have been a haemophiliac and her mother a carrier. As the gene has a very low frequency in the population, the chance of this happening is very remote.
       (iii) The gene for coat colour is carried on the X chromosome. The female is homogametic, and the genes for yellow and black are co-dominant.

♀	*BB*	*BY*	*YY*
	black	tortoiseshell	yellow
♂	*B–*		*Y–*
	black		yellow

35 (i) ♀ 6 black, long-tailed
          ♀ 2 black, short-tailed
          ♂ 6 white, long-tailed
          ♂ 2 white, short-tailed

(ii) ♀ 4 black, long-tailed
     ♀ 4 white, long-tailed
     ♂ 4 black, long-tailed
     ♂ 4 white, long-tailed
(iii) ♀ 3 black, long-tailed
     ♀ 1 black, short-tailed
     ♀ 3 white, long-tailed
     ♀ 1 white, short-tailed
     ♂ 3 black, long-tailed
     ♂ 1 black, short-tailed
     ♂ 3 white, long-tailed
     ♂ 1 white, short-tailed

36 (i) All 75 mm
(ii) 1 of 39 mm: 4 of 48 mm: 6 of 57 mm: 4 of 66 mm: 1 of 75 mm:
(iii) 8 of 48 mm: 8 of 57 mm
(iv) All 39 mm

37 (a) The expected ratio, if Mendel's laws are obeyed, is 1 : 1 : 1 : 1. Significant departure from this ratio indicates that the genes are on the same chromosome, and the cross-over value 8/238 × 100 = 3% shows that the genes are closely linked on the chromosome.
(b)

E	C	D	F	G	A	B
	7.5	6.3	5.2	1.6	3.4	4.6

11 The Hardy–Weinberg Law, Evolution and Classification

1 (a) **2** (b) **3** (e) **4** (b) **5** (b) **6** (d)
7 (b) **8** (a) **9** (e) **10** (b) **11** (a) **12** (e)
13 (d) **14** (c) **15** (d) **16** (d) **17** (c)
18 (b) **19** (c).

20 (a) See Table 54. and Fig. 128.

Table 54

Population	Genotype frequency		
	AA	*Aa*	*aa*
1	1.00	0.00	0.00
2	0.64	0.32	0.04
3	0.36	0.48	0.16
4	0.16	0.48	0.36
5	0.04	0.32	0.64
6	0.00	0.00	1.00

(b) If the number of recessive individuals (*aa*) in the population was known, the graph could be used to determine the number of homozygous dominant (*AA*) and heterozygous (*Aa*) individuals in the population.

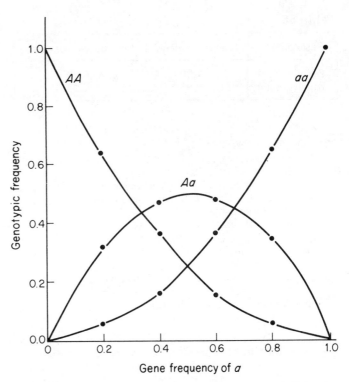

Fig. 128 The relationship between the frequency of allele *a* in the population and the frequency of genotypes *AA*, *Aa* and *aa*

(c) See Table 55.

Table 55

Mating	Product	Expected frequency
Black ×black	0.9 ×0.7	0.63
Black ×white	0.9 ×0.3	0.27
White ×black	0.1 ×0.7	0.07
White ×white	0.1 ×0.3	0.03
		1.00

(d) (i) *AA* 0.64   *Aa* 0.32   *aa* 0.04
   (ii) *AA* 0.64   *Aa* 0.32   *aa* 0.04
   (iii) *AA* 0.36   *Aa* 0.48   *aa* 0.16
   (iv) *AA* 0.6725   *Aa* 0.295   *aa* 0.0325
   (v) Population (i).
(e) Mutation, migration, natural selection, genetic drift
(f) $q^2 = 0.00005$ $q = 0.0071$ and $p = 0.9929$
   $2pq = 1.41\%$ of population

21 (a) $b = 0.1$ and $B = 0.9$
(b) $T = 0.42$ and $t = 0.58$
(c) Providing there has been no selection of genotypes before the elvers reach the coast of Europe, application of the Hardy–Weinberg Law suggests that European and American eels are derived from different spawning populations (Table 56).

On the other hand, if forces of selection are in operation, it may be that genotypes *Aa* and *aa* are eliminated from the population before elvers reach the coast of Europe.

Table 56

Population	Frequency gene A	Frequency gene a
European eel	1.00	0.00
American eel	0.95	0.05

22 (a) Darwin, Wallace
    (b) Predators, disease
    (c) Bees, banded snails
    (d) Monarch butterflies and milkweeds, termites and fungi
    (e) Darwin's finches, limb structure in mammals
    (f) Dodo, Tasmanian wolf
    (g) Kangaroo, wallaby
    (h) *Sphenodon, Ginkgo biloba*
    (i) Mule, loganberry
    (j) *Peripatus, Homo erectus*
    (k) Whales and fish, desert succulent plants
    (l) Wings of bat and bird, claws and hooves
    (m) Molar teeth, appendix of man

23 (a) (i) Discontinuous variation (ii) Continuous variation
    (b) (i) Ear length appears to be controlled by a single pair of genes, which are incompletely dominant: AA = short, Aa = intermediate length, aa = long.
        (ii) Tail length appears to be controlled by the interaction of several genes.
    (c) Ear-length
    (d) It would be necessary to know the total number of short-eared or long-eared mice in the population.
    (e) Stabilising selection eliminates small numbers of extreme individuals from both ends of the spectrum of variation.
        Directional selection eliminates small numbers of extreme individuals from one end of the spectrum of variation.
        Disruptive selection eliminates large numbers of intermediate forms, leaving the population composed of extreme types at each end of the spectrum of variation.
    (f) Stabilising selection would have eliminated most short-eared and long-eared mice from the population, leaving a population composed largely of mice with ears of intermediate length.
        Directional selection would have reduced the numbers of short-eared and long-eared mice in the population.
        Disruptive selection would have eliminated those mice with ears of intermediate length, leaving the population composed of short-eared and long-eared types only.
    (g) Disease, shortage of food, injury, winter cold, summer drought
    (h) Assuming that wood mice have the ability to dig to the floor of the soil layer, and could nest at any depth in the soil, then
        (i) predators might eliminate those individuals who nested on or close to the surface,
        (ii) fire would burn out nests that were on, or close to the surface, especially if dried grass had been trailed along the burrow,
        (iii) deep-nesting mice are using the wettest region of the soil, and might lose their young as a result of flooding.
        These factors, acting together, probably account for the largest number of nests being present in the central region of the soil layer.

24 (a) A mutation is a change in the structure of a gene, or chromosome, that results in the appearance of a new allele within a population.
    (b) (i) Interchange translocation (ii) Deletion, or deficiency (iii) Duplication (iv) Inversion
    (c) (i) Polyploidy (ii) Point mutation (substitution)
    (d) (i) Mustard gas, formaldehyde, hydrogen peroxide
        (ii) X-ray, UV light, gamma radiation
    (e) The DNA code often consists of more than one triplet of bases coding for the same amino acid. Providing the changed DNA sequence coded for an identical amino acid, there would be no change in the identity of amino acids in the polypeptide chain.
    (f) One allele   (ii) Heterozygous
    (g) (i) Genes may mutate to either dominant or recessive form.
        (ii) Most genes mutate to recessive form.
    (h) Mutation (i) would be detrimental. Seedlings of this type would soon die, as they could not photosynthesise without chlorophyll.
        Mutation (ii) could be beneficial. If three cotyledons provide a larger leaf area in which photosynthesis can take place, mutant seedlings might grow more rapidly than normal seedlings, and thereby gain some selective advantage over them.

25 (a) Over a period of 24 hours, acriflavine diffuses from the well into the surrounding agar, forming a diffusion zone in which concentrations are highest at the centre and lowest at the margins of the circular zone surrounding the well. In zone A concentrations of acriflavine are lethal to the yeast cells. All of the yeast cells within this zone have been killed and therefore fail to produce any colonies. In zone B concentrations of acriflavine are lower. At these concentrations some yeast cells are killed, but others suffer chromosome damage, leading to the development of mutant forms, which include petites and giants. Zone C, however, lies outside the diffusion zone of acriflavine. Hence, the yeast cells in this zone survive and each produces a normal colony.
    (b) Acriflavine is a toxic compound, lethal at high concentrations and mutagenic at lower concentrations, causing damage to chromosomes.
    (c) The diffusion zone of acriflavine would cover a larger area, as the rate of diffusion is temperature-dependent. Zone A would probably be broader, zone B be located nearer to the edge of the plate, and zone C narrower, with fewer normal colonies. As high temperature is an additional factor capable of causing mutations, the number of mutant colonies would probably be increased.
    (d) Number of normal colonies = 12
        Number of petite colonies = 4
        The circumference of the dish $= \pi d = 28.26$
        Approximate no. normal colonies $= \dfrac{28.26}{3.0} \times 12 = 113$
        Approximate no. petite colonies = 38

26 (i) *Over-reproduction.* Most organisms produce more offspring than will survive to maturity. Any organism, if all of its progeny survived, could eventually colonise large areas of terrestrial or aquatic habitats, occupying all of the space capable of supporting them.
    (ii) *Struggle for existence.* The fact that many offspring die before reaching maturity, and that few populations are undergoing rapid expansion, implies a struggle for exist-

ence between species and between individual members of a species, for available food, space etc.

(iii) *Natural selection.* All organisms must survive against the forces of natural selection, such as predators, parasites, cold, heat, drought, flood, fire, wind etc., all of which are capable of eliminating those that are unfit.

(iv) *Survival of the fittest.* One important outcome of the struggle for existence is survival of the fittest; that is, survival of the best adapted individuals, while poorly adapted individuals die prematurely. Those that survive pass on many of their beneficial traits to their offspring which, in their turn, must continue the struggle against their contemporaries.

(v) *Isolation.* Geographical isolation of a small breeding population, such as the inhabitants of an island, cut off from larger populations by water, mountain ranges etc. for a considerable period of time, may evolve into a new species; that is, one which cannot breed with members of the larger, non-isolated population.

(vi) *Origin of a new species.* Darwin envisaged the new species arising from individuals that had been subjected to natural selection and isolation.

(vii) *Gene mutation.* Neo-Darwinists view the raw material of evolution as the new alleles, or traits, introduced into the population as a result of mutations of genes or chromosomes.

(viii) *Genetic recombination.* Variety is increased, and spread throughout the population when these mutations are spread, and recombine with other genes in the gene pool.

(ix) *Natural selection.* The forces of natural selection act on the gene pool, selecting certain genotypes and either reducing them in the population, or eliminating them from the population.

(x) *Reproductive isolation.* Any form of reproductive isolation, ecological or behavioural, as well as geographical, can cause more inbreeding, which results in an increase in the number of homozygous individuals, at the expense of heterozygotes. The homozygotes express more of the alleles in the population, especially the recessive alleles. Therefore, the rate of change in the population is accelerated.

(xi) *Origin of a new species.* A new species arises when the gene pool of the inbreeding population is incompatible with that of the main population.

27 (a) Model A is monophylectic, which implies that all life forms have originated from a single common ancestor. Additionally, the model implies linear descent, a single species having been formed from each ancestral form. Model B is polyphylectic, which implies that existing forms may have been derived from more than one common ancestor. Secondly, the model implies divergent evolution, with some species evolving into two or more forms.

(b) (i) Artificial selection of animals by man illustrated that the appearance of an organism, such as a horse or pigeon, could be changed over a relatively short period of time, providing certain characters were selected generation after generation.

(ii) The entire fauna of the Galapagos Islands resembled that of the South American mainland. Even so, populations on the Galapagos Islands were sufficiently divergent, and had been isolated for a sufficient period of time, to allow species formation to have taken place.

(iii) Darwin's finches illustrated adaptive radiation. On islands where the only available source of food was located beneath stones, in the bark of trees, and in

succulent parts of cacti, early colonisers had to learn how to obtain food. Different sub-populations arose, each feeding on a different food source. Natural selection operated in favour of forms of bill which best adapted different populations to particular methods of feeding. In the course of time, and following from isolation, separate populations arose, each with a type of bill adapted for feeding on a particular diet.

(c) (i) The chief selective pressure on populations of *Cepaea* is exerted by thrushes, which remove snails from the colony, crack their shells on stones, and eat the soft parts. The selective advantage of different shell types is partly a function of the background on which they occur. Thus, against the dark brown leaf litter of the woodland floor, dark brown shells are least conspicuous to the thrushes. Similarly, green-yellow shells are well camouflaged in grassland, and a banded pattern blends in well with a background of erect, dead stems and leaves, found in sward.

(ii) Early in the spring, and again during the winter, when the woodland floor was predominantly brown, the yellow shells were relatively conspicuous and therefore at a selective disadvantage. As the season progressed, and the background became greener, this disadvantage lessened, so that the yellow shells became selectively neutral. During the summer, however, yellow-green shells were probably at a selective advantage.

(iii) Experiments with *Cepaea nemoralis* have indicated that unbanded individuals are more heat-resistant than banded individuals. Yellow snails are more resistant to cold than brown snails or banded snails. These and similar characteristics indicate that colour and banding are subject to strong non-visual selection because they are associated with important physiological advantages.

28 (a) (i) Monera   (ii) Protozoa   (iii) Sarcodina
        (iv) Basidiomycetes   (v) Chlorophyta
        (vi) Bryophyta   (vii) Tracheophyta
        (viii) Angiospermae   (ix) Monocotyledonae
        (x) Animalia   (xi) Arthropoda   (xii) Crustacea
        (xiii) Insecta   (xiv) Arachnida   (xv) Diplopoda

    (b) Mosses: ii, iv, vii, viii, ix
         Ferns: i, iii, v, vi, x

    (c) Amphibians: i, ii, iv, viii, ix, x
         Reptiles: iii, v, vi, vii

29 *Table 47* can be completed with the following answers.
    (i) Mammalia: Aves: Amphibia: Pisces
    (ii) Badger, fox: blackbird, thrush: frog, toad: roach, perch
    (iii) Dry, covered by hairs: dry scaly on legs, otherwise covered by feathers: moist, naked, used in respiration: moist, covered by scales
    (iv) Internal: internal: external: external
    (v) Embryo protected within embryonic membranes, fed via placenta: embryo protected within embryonic membranes, enclosed by an egg shell: embryo surrounded by gelatinous layer: egg surrounded by gelatinous layer
    (vi) Lungs: lungs and air sacs: skin and lungs: gills
    Table 48 can be completed with the following answers.
    (i) Animalia: Plantae: Fungi
    (ii) Arthropoda: Tracheophyta:—
    (iii) Insecta: Angiospermae: Zygomycetes
    (iv) Holozoic: holophytic: saprophytic
    (v) None: stem tubers: asexual spores in sporangia
    (vi) None: cellulose: chitin